Methods of Statistical Analysis of Fieldwork Data

Methods of Statistical Analysis of Fieldwork Data

Peter St John
and Dave Richardson

THE GEOGRAPHICAL ASSOCIATION

ISBN 1 899085 16 5
First published 1990
First revised edition 1996
Impression number 10 9 8 7 6 5 4 3
Year 2000

Published by the Geographical Association, 160 Solly Street, Sheffield S1 4BF.
The Geographical Association is a registered charity: no. 313129

The Publications Officer of the GA would be happy to hear from other potential authors who have ideas for geography books. You may contact the Publications Officer via the Geographical Association at the address above. The views expressed in this publication are those of the authors and do not necessarily represent those of the Geographical Association.

Acknowledgements

The authors would like to thank the following people for their help with this book:

Sixth-form groups in Lancashire, for identifying the need for a manual of this nature

Mr E E Jones, former Warden of Lancashire Field Study Centre, Hothersall Lodge, for his encouragement and support

Mrs E Kenny, for patiently and expertly typing and retyping (and retyping)

Mr J Fallas, for the cartoon illustrations

The cover photograph and frontispiece were taken by Dave Richardson

Diagrams: Paul Coles and Jason Haigh

Design and typesetting: White Line Publishing Services

Cover design: Chris Hand

Printed and bound in England by The Thanet Press

CONTENTS

INTRODUCTION

In modern fieldwork much of the information is collected in the form of numerical quantities. Often the purpose is to compare or contrast sets of data and therefore it may be necessary to show that there is a real similarity or difference between them. It is often the case that sets of information are not clearly distinguishable and so the use of an appropriate statistical method is necessary to identify similarities and differences.

The purpose of this book is to help you to choose the appropriate statistical method to apply to your data and also how to conduct the chosen test. It is essentially a 'recipe book' and does not aim to derive or explain the derivation of any mathematical formulae.

The tests are set out in a relatively standardised format, i.e.

- Introduction/title
- When to use
- Examples of calculation
- Cautionary points on application of test and analysis of results.

This booklet is aimed at A-level geographers and biologists who are required to apply statistical tests to validate their analysis of data but who are not necessarily expected to be first-class mathematicians.

Using the book

It is not intended that this book is read cover to cover but as a manual. However, the first two chapters — 'Design of a Study' and 'Background to Statistical Methods' — should be consulted before choice of project and statistical analysis is made. The section below and the flowchart on pages 8–9 will refer the reader to the required statistical method.

Choosing the appropriate method

Having identified your intentions, use the chart on pages 8–9 before you collect your data. Note carefully the form in which the data must be collected and any limitations of sample size. Note also the nature of the data's distribution. This chart could save a lot of time that may be wasted collecting inappropriate data.

INTENTION

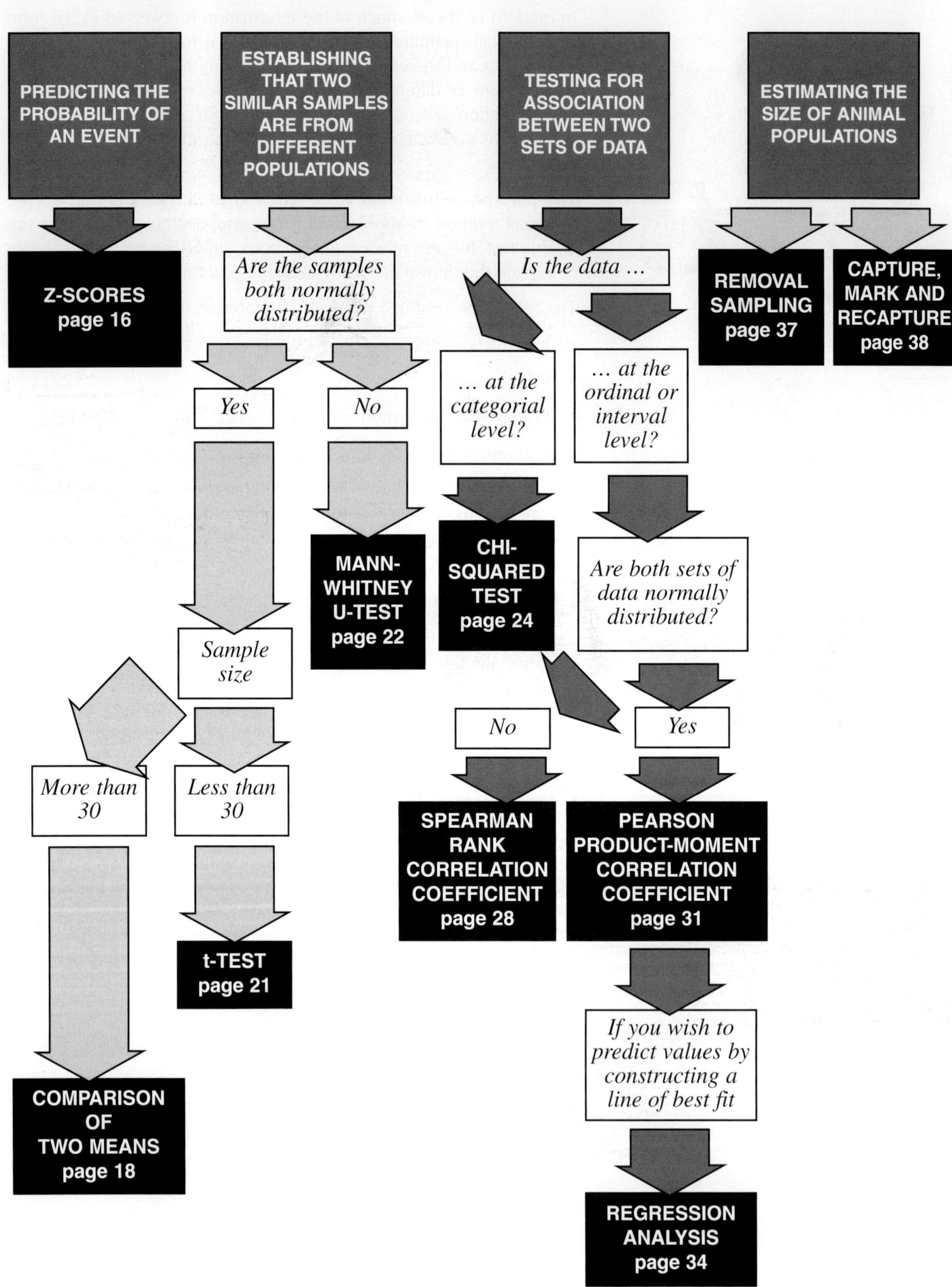

Flowchart for choosing the appropriate statistical method

INTENTION

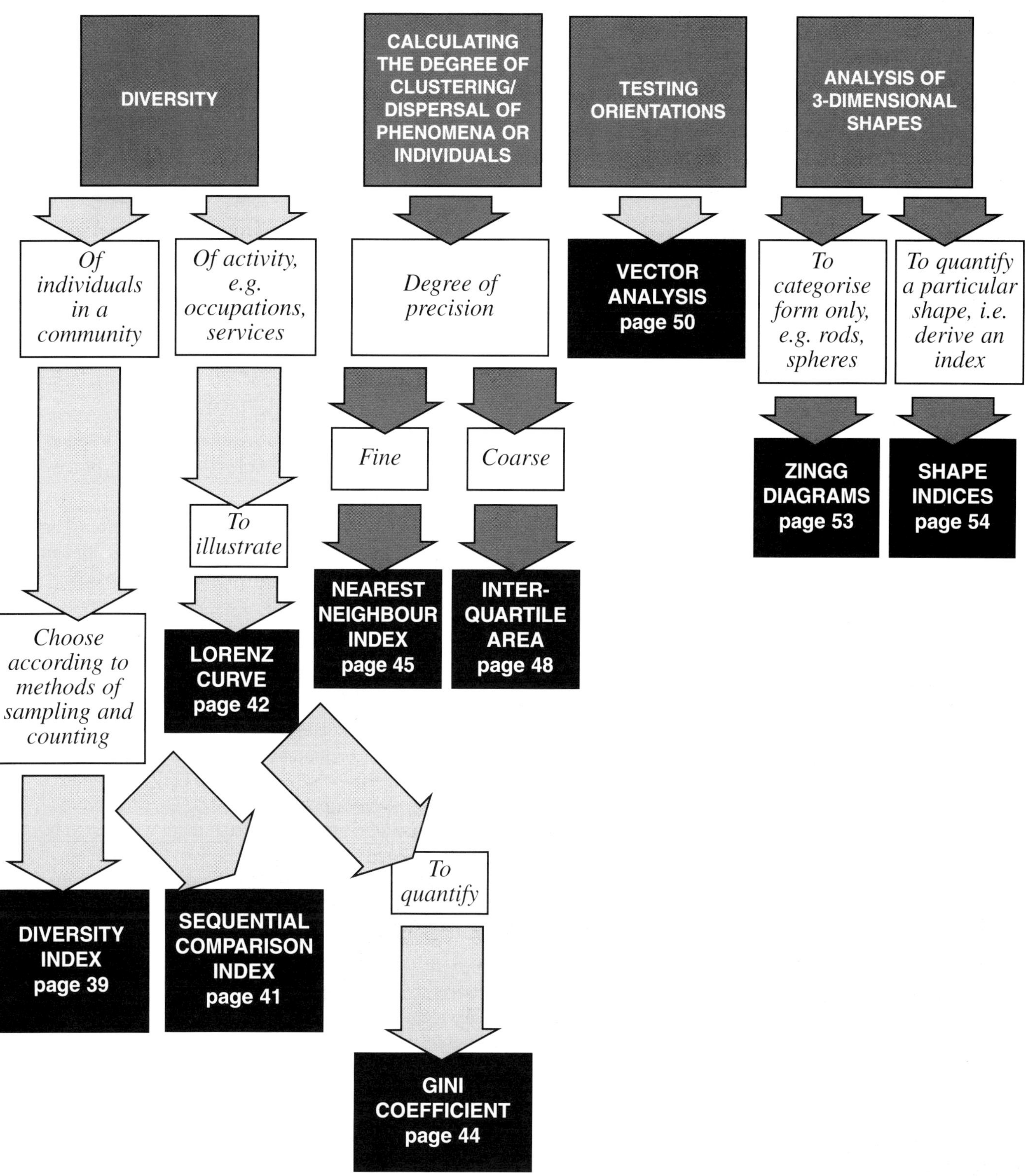

DIVERSITY
CALCULATING THE DEGREE OF CLUSTERING/ DISPERSAL OF PHENOMENA OR INDIVIDUALS
TESTING ORIENTATIONS
ANALYSIS OF 3-DIMENSIONAL SHAPES
Of individuals in a community
Of activity, e.g. occupations, services
Degree of precision
VECTOR ANALYSIS page 50
To categorise form only, e.g. rods, spheres
To quantify a particular shape, i.e. derive an index
Fine
Coarse
ZINGG DIAGRAMS page 53
SHAPE INDICES page 54
To illustrate
NEAREST NEIGHBOUR INDEX page 45
INTER-QUARTILE AREA page 48
Choose according to methods of sampling and counting
LORENZ CURVE page 42
To quantify
DIVERSITY INDEX page 39
SEQUENTIAL COMPARISON INDEX page 41
GINI COEFFICIENT page 44

1: DESIGN OF A STUDY

Different statistical techniques are appropriate in different circumstances, and may only be applied to data collected in a particular way. Therefore, when designing a study, it is vital that thought be given in the initial stages to the manner in which the data is collected so that it can be processed later by the appropriate statistical method.

Generally a study begins with an observation. If a difference or trend appears possible a **hypothesis** (idea) may be formed. A hypothesis is an idea or tentative theory not supported by hard evidence usually stating that some relationship between variables exists. The hypothesis may then be tested using an appropriate method to collect sufficient information.

On the basis of the analysis of the information, the hypothesis:

1 may be accepted

2 may be rejected

3 may be inconclusive and more information required.

If the original hypothesis is rejected, then a new one may be inserted, more (and possibly different) information collected and the process repeated.

This process is called the **scientific method** (see Figure 1.1). It must be emphasised that if statistical treatment is necessary then this is applied to the data and on the basis of the results the hypothesis may be accepted or rejected. However, it is not always necessary to use statistics to support an idea.

The null hypothesis

At this point it may be useful to explain the concept of the **null hypothesis**. A hypothesis is an idea or tentative theory not supported by hard evidence, usually stating that some relationship between variables exists. The null hypothesis is the opposite statement, i.e., that no relationship exists between variables, and it is this statement which is to be tested. Therefore, rather than prove the hypothesis we disprove the null hypothesis. This may seem cumbersome but it does mean a more impartial approach so that we do not see relationships just because we want to.

For example, we may have observed that the pebbles in one part of a river appear to be larger than in another part of the river. Being cautious we would tentatively state that there is **no** difference (our null hypothesis) until experimental evidence shows otherwise. It is, therefore, the null hypothesis that we are testing. If we show to statistical satisfaction (see 'significance and confidence limits' in Chapter 3) that we cannot accept our null hypothesis then it is rejected and an alternative hypothesis is substituted. In our example, which is shown on pages 18–20, the hypothesis that there is a significant difference in stone sizes at our two sites was substituted.

The null hypothesis is often printed as H_0. The alternative hypothesis is referred to as H_1.

Illustration of the scientific method

Figure 1.1 (below) shows an outline of the scientific method along with an example of its application. The study should begin with an **observation,** e.g. two different soils are found in two different areas of gritstone moorland. The broad problem has been identified. Why is this so? The **analysis of the problem** has begun. Which factors are important in soil formation?

An initial **description** of the soils is required comparing the characteristics of the two. Features including pH, organic matter, water content, mineral content, texture and appearance may be recorded and compared. It might be noted that the organic content of peat is very much higher than that of podsol and that this may be worthy of further investigation (Table 1.1).

	Peat	Podsol
pH	4.5	5.5
% Water	93	62
% Organic matter	87	12
% Mineral content	11	81

Table 1.1: Some characteristics of a peat and a podsol compound

So a **choice of specific problem** has been made. Why does the organic content vary so much? Is it possibly due to different rates of decomposition in the soil?

It is possible at this point to **formulate a hypothesis**. Great care is needed with choice of hypotheses. The best are those which are clear and specific.

Figure 1.1: An outline of the scientific method with illustration

Stage	Illustration
1 Observation	The soil on the flat top of a moorland hill (peat) is different from that found on the steep sides (podsol)
2 Analysis of problem	Why are the soils different? Which factors and processes define the nature of the soil?
3 Description	Summary of factors, processes, and identifiable characteristics of that soil.
4 Choice of specific problem	The rate of decomposition of humus in the soils at different sites is important in creating a largely organic soil on the top and a mainly inorganic soil on the sides.
5 Formation of hypothesis	The rate of decomposition of dead plant material differs in the two soils thereby affecting the nature of each soil.
6 Experiment	Cotton (cellulose) thread is buried at a constant depth below the surface of the soil and left to decompose for ten weeks.
7 Results	The breaking strain of cotton strands is measured and recorded as a percentage of its original breaking strain.
8 Analysis of results	When the individual breaking strains for each cotton thread is plotted against frequency there appears to be a difference between those threads from the peat and those from the podsol although there is some overlap. Is there a real difference? Apply the appropriate statistical test. (See Mann-Whitney U-test on pages 22–23 for treatment of these results.)

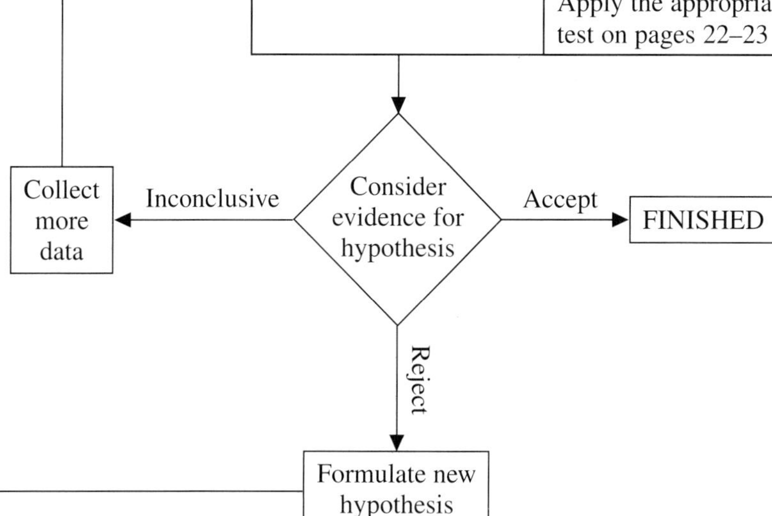

Many projects founder because the student initially collects as much data as possible and then looks for a hypothesis to fit the results. At best much of the information collected may be useless, or at worst the study may become a vague collection of loosely related information.

The hypothesis chosen for the example is: 'The rate of decomposition of dead plant material differs between a peat and a podsol, thereby affecting the nature of each soil'.

This is specific and leads us to look for a method of testing the hypothesis. It is here that forward planning is required. The following questions should be asked:

- Is relevant experimental work possible?
- Is equipment available?
- How time-consuming will the experimental work be?
- Are suitable sites easily accessible?

If the answers to these questions are satisfactory then the **experimental collection of data** may be planned. Again forward thinking is required if statistical analysis is required. For analysis to be possible the data will have to be in an appropriate form. As with hypothesis formulation, the potential for collection of worthless data is great, as is the accompanying waste of time. For assistance here consult the flowchart for choosing the appropriate statistical method on pages 8–9. Pay special note to the amount of data required for valid results. Too little or too much represents another waste of time.

The experiment designed to measure rates of cellulose decomposition involves burying ten strands of cotton (not man-made fibres, e.g. acrylic) in two bundles of five each in the peat at a depth of 5 cm and repeating this in the podsol. These are left for ten weeks and recovered. The breaking strain of each strand of cotton is found by hanging weights on it until the cotton breaks. The original breaking strain of the cotton is found in the same way using cotton that has not been buried, and the breaking point of the buried cotton expressed as a percentage reduction in original breaking strain. The experiment was designed so that the Mann-Whitney U-test may be used if statistical analysis is required.

The results from this experiment are used to illustrate the Mann-Whitney U-test and this may be followed to its logical conclusion on pages 22–23.

2: BACKGROUND TO STATISTICAL METHODS

Before embarking on the application of statistical analysis of your data some knowledge of the common terms and ideas is required.

Assume that you had measured the lengths of ten flowering stalks of the soft rush (*Juncus effusus*) in centimetres. These were:

72	84	69	93	90
76	78	67	74	81

The lengths are not all identical (we would not expect them to be) but a useful piece of information would be their average or **mean** length. This is calculated by summing all the individual lengths and dividing by the number in the sample. The mean length of the rushes is obtained by Equation 2.1.

$$\text{Mean} = \frac{\text{Total of all rush lengths}}{\text{number in sample}}$$

$$= \frac{784}{10}\text{ cm} \quad = \mathbf{78.4\ cm}$$

Equation 2.1: Mean

Conventionally the mean is expressed as $\bar{x}$ (pronounced x-bar), the number in the sample as n, and the total of all individuals in the sample as Σx. The Greek letter Σ (sigma) is used to denote 'the sum of'. Hence Σx means the sum of all the individuals in the sample. Therefore Equation 2.1 becomes:

$$\bar{x} = \frac{\Sigma x}{n}$$

Another useful statistic concerning our data is the **median**. This is that value for which half of the values lie above and half lie below. A line-plot is an easy method of establishing the median (Figure 2.1).

Figure 2.1: Lineplot showing distribution of rush lengths

As there are ten measurements, the median is situated so that there are five measurements above and five below the value, i.e. mid-way between the fifth and sixth values.

Fifth value = 76 cm

Sixth value = 78 cm

$$\text{Therefore median} = \frac{78+76}{2} \quad = \mathbf{77\ cm}$$

Equation 2.2: Median

The mean and the median values are different in this case (78.4 and 77 respectively), and with field data this is usual. **The larger the difference, the more a set of data is said to be skewed.**

If all the sample lengths clustered equally around the mean they could be plotted as in Figure 2.2. The mean value would be at the apex of the curve because values are clustered equally either side. The median value would be at the same point because there are equal numbers of samples either side of the apex. **In this case mean = median.**

Figure 2.2: A symmetrical curve

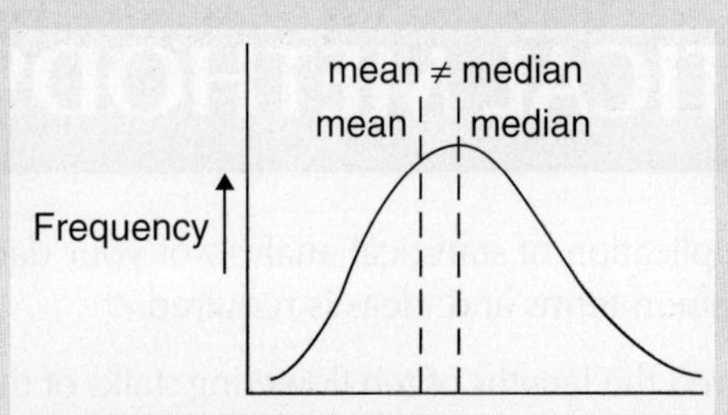

Figure 2.3: A skewed curve

If the curve were skewed it would look like Figure 2.3. The difference between the values for the mean and the median indicates the degree to which that curve is skewed, i.e. in our example 78.4 – 77 = 1.4.

Standard deviation and the normal curve

If a set of continuous variables is plotted the result is likely to be a bell-shaped curve (this is called the **normal curve**), e.g. if the lengths of rush stems are plotted against frequency (Figure 2.4).

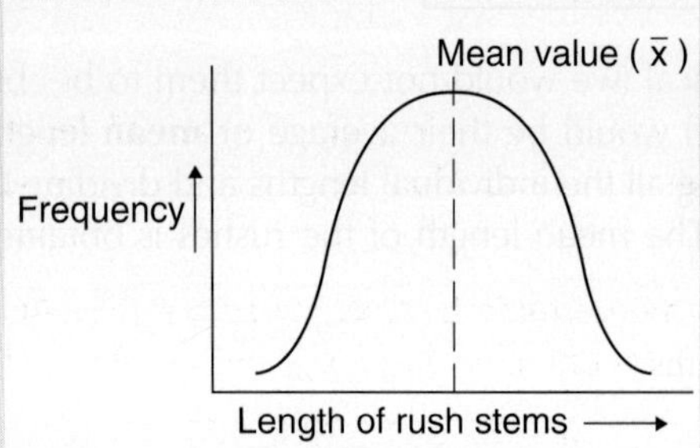

Figure 2.4: A continuous variable plotted against frequency

The curve implies that most individuals are aggregated around the average or mean length but increasingly fewer are very long or very short. In theory the mean would correspond to the middle of the curve around which each side is symmetrical. The ideal curve has certain fixed mathematical properties. It must be emphasised that information collected in the field will only approximate to the ideal curve when plotted but that the larger the sample the better will be the approximation. (A good argument for advocating as large a sample as is practicable.) The relevance to those of us who employ statistics as a tool is that we can use the properties of the normal curve to assess the reliability of the conclusions from our data.

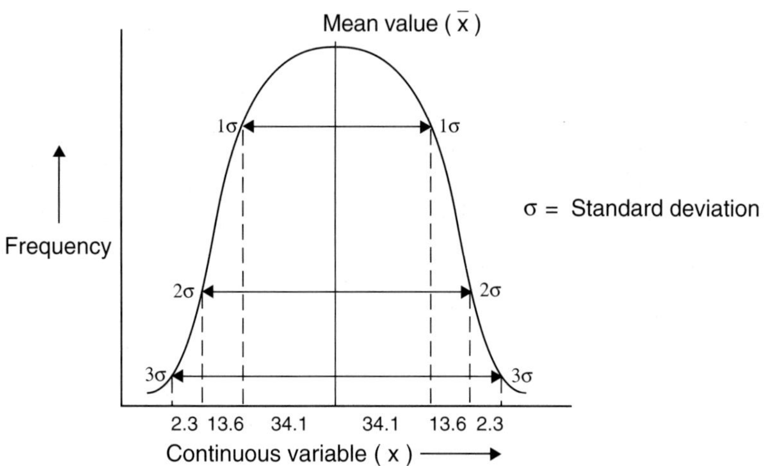

Figure 2.5: The ideal normal curve and the proportions of individuals scattered at distances from the mean

In a normal curve 68.2% of all individuals will lie within one **standard deviation** (σ) either side of the mean (see Figure 2.5). Similarly 95.4% of all individuals are clustered within 2σ of the mean. The standard deviation is a measure of the dispersal of observations around the highest point (which is the mean in a normal curve). If σ is small then the population is tightly packed around the mean, and conversely if σ is large then the curve is flatter and the observations spread further from the mean (see Figure 2.6).

Note: That the distribution of items from the mean is predictable is a vital cornerstone of statistical theory.

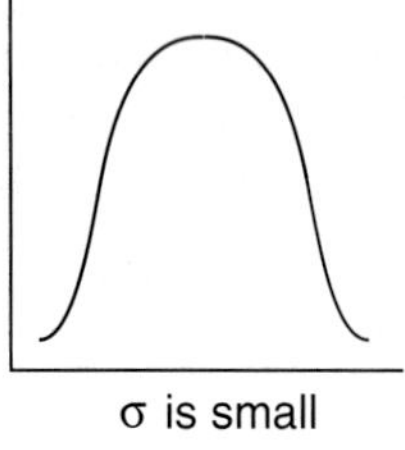

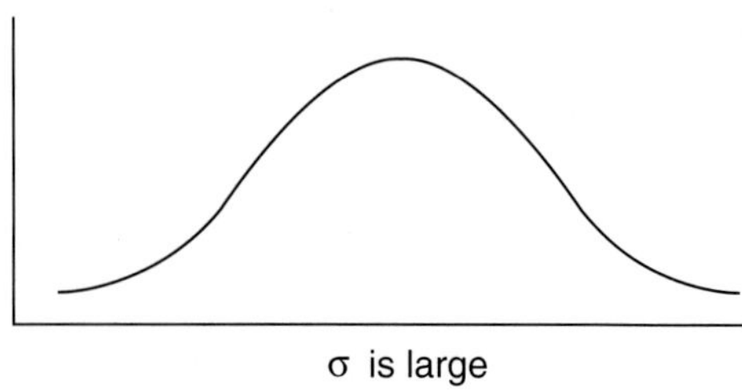

Figure 2.6: The relationship between standard deviation (σ) and the shape of a normal curve

Calculation of the standard deviation

The standard deviation (σ) is calculated by the equation

Equation 2.3

$$\sigma = \sqrt{\frac{\Sigma x^2}{n} - \bar{x}^2}$$

where x is the variable, $\bar{x}$ is the mean of the variable and n is the total number in the sample.

When to use

The standard deviation of a set of values distributed normally will assess the reliability of the mean. It can also be used to predict the probability of any stated event occurring (see 'Z-scores' on page 16).

Example 1

To calculate the mean and standard deviation of a set of results which determine the oxygen concentration of a sample of pond water

Using a chemical technique to determine the oxygen content (mg/l) of pond water the group of students obtained the following results:

2.9	3.0	3.7	3.1	4.2	2.9	3.3	3.4
3.6	3.5	3.3	3.9	3.2	3.7	3.5	3.6

Clearly as each student was testing the same sample of pond water with the same chemicals the variation must be due to sampling error. Some sampling error is inherent in all fieldwork results but this does not prevent us from using the results. The question is, what is the oxygen concentration of this pond water and how reliable is our estimate?

The first task is to set the data out in a table as in Table 2.1 (left), then to calculate the outcome (see Equations 2.4 and 2.5).

x	x^2
2.9	8.41
3.0	9.00
3.7	13.69
3.1	9.61
4.2	17.64
2.9	8.41
3.3	10.89
3.4	11.56
3.6	12.96
3.5	12.25
3.3	10.89
3.9	15.21
3.2	10.24
3.7	13.69
3.5	12.25
3.6	12.96
$\Sigma x = 54.8$	$\Sigma x^2 = 189.66$

Table 2.1: Data from Example 1

Equation 2.4

$$n = 16$$

$$\bar{x} = \frac{54.8}{16} = 3.43$$

$$\bar{x}^2 = 11.73$$

$$\text{Using } \sigma = \sqrt{\frac{\Sigma x^2}{n} - \bar{x}^2}$$

$$\sigma = \sqrt{\frac{189.66}{16} - 11.73}$$

Equation 2.5

$$\sigma = \mathbf{0.35}$$

Analysis

The analysis of our data shows that the mean value for the oxygen concentration is 3.43 and that the standard deviation is 0.35.

Referring to our theoretical normal curve we can now say that there is:

- a 68% probability of the mean lying in the range 3.43 ± 0.35 (σ) i.e. between 3.78 and 3.08.
- a 95% probability of the mean lying in the range 3.43 ± 0.7 (2σ) i.e. between 2.73 and 4.13.
- a 99.5% probability of the mean lying in the range 3.43 ± 1.05 (3σ) i.e. between 2.38 and 4.48.

Clearly we now have a better idea of the true oxygen concentration of the pond water.

x	x^2
4.3	18.49
1.5	2.25
12.1	146.41
10.1	102.01
8.7	75.69
8.4	70.56
6.3	39.69
8.1	65.61
9.0	81.00
7.2	51.84
$\Sigma x = 75.7$	$\Sigma x^2 = 653.55$

Table 2.2: Data from Example 2

Example 2

To calculate the standard deviation of rainfall figures for a Lancashire coastal resort in August over the last ten years

The rainfall figures for August over the last ten years at a Lancashire resort are as follows:

4.3	1.5	12.1	10.1	8.7
8.4	6.3	8.1	9.0	7.2

The procedure is as in Example 1. First tabulate the data: Table 2.2 (left), then calculate the outcome (see Equations 2.6 and 2.7).

$$n = 10$$

$$\bar{x} = \frac{75.7}{10} = 7.57$$

Equation 2.6

$$\bar{x}^2 = 57.30$$

$$\text{Using } \sigma = \sqrt{\frac{\Sigma x^2}{n} - \bar{x}^2}$$

$$\sigma = \sqrt{\frac{653.55}{10} - 57.30}$$

Equation 2.7

$$\sigma = \mathbf{2.84}$$

The mean August rainfall is 7.57 cm and the standard deviation is 2.85.

Therefore there is a 95% probability that there will be 7.57 ± 5.68 cm of rain in August (i.e. 2 standard deviations from the mean). This means there is a 95% probability that between 1.89 and 13.25 cm of rain will fall in August.

Z-scores

When to use

By developing the idea of the standard deviation a little further it is possible to assess the probability of any stated occurrence using the Z-score. It is also possible to assess the probability of an occurrence over a stated range of values. The Z-score statistic has a predictive value which may be of use with certain types of information.

Example 1
(using the rainfall data in the previous example)

What is the probability of a completely dry August at the Lancashire resort?

The probability of any given quantity of rain can be calculated using the following formula:

Equation 2.8

$$Z = \frac{x - \bar{x}}{\sigma}$$

where x is the stated value, $\bar{x}$ is the mean and σ is the standard deviation.

For this example, $x = 0$, $\bar{x} = 7.57$ and $\sigma = 2.84$. Using the Z-score equation,

$$Z = \frac{0 - 7.57}{2.84}$$

Equation 2.9

$$= \mathbf{-2.67}$$

This means that the stated value of 0 lies 2.67 standard deviations away from the mean. Refer to Appendix 1, column B (page 56) for the probability of this occurrence.

Note on expressing probability

The probability of an occurrence is expressed as a number between 0 and 1. A probability of 0 means that the event will never happen; a probability of 1 means that it is certain to occur. To convert a probability into a percentage simply multiply it by 100, e.g. a probability of 0.5 becomes 50%.

The probability is 0.004 or **0.4%** (i.e. 4 times per thousand). Therefore a completely dry August is experienced once every 250 years on average! (This is why people go to Spain for their holidays.)

The Z-score can also be used to assess the probability of an event occurring between two values.

Example 2

What is the probability of between 6 and 8 cm of rain falling in August?

Method

Calculate the Z-score for each stated value.

Equation 2.10

$$\text{For } x = 8 \quad Z = \frac{8 - 7.57}{2.84} = \mathbf{0.15}$$

Equation 2.11

$$\text{For } x = 6 \quad Z = \frac{6 - 7.57}{2.84} = \mathbf{-0.55}$$

Refer to the table of Z-scores (Appendix 1, column A, page 56) to find the probability of a value lying between the mean and each given value of Z.

$-0.55 \rightarrow 0.226$

$+0.15 \rightarrow 0.079$

Add these values together:

$0.226 + 0.079 = 0.305$

The probability of between 6 cm and 8 cm of rain falling in August is 0.305 or 30.5%.

Alternatively, between 6 and 8 cm of rain will fall in August in about one year in three.

3: COMPARING SAMPLES FOR DIFFERENCE

The comparison of two means

The problem to be solved here concerns the comparison of two sets of data to ensure that they are actually different. If we plotted the data on to a graph (Figure 3.1) we might observe:

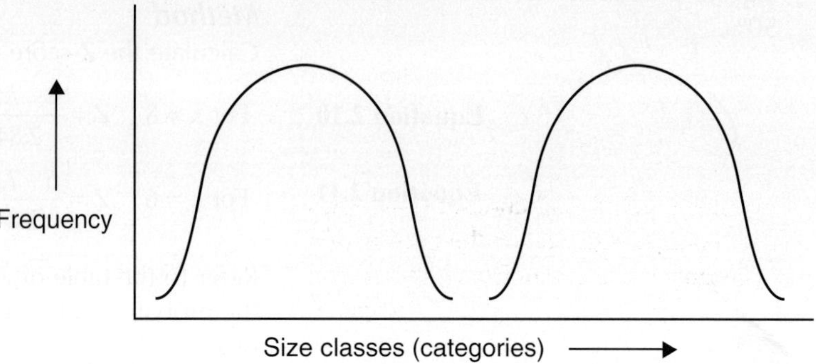

Figure 3.1: Ideal normal curves for two sets of data where the standard deviations are the same but the means are obviously different

Clearly there are two separate sets (populations) here with the same standard deviations but quite different means.

More commonly we might see a graph resembling Figure 3.2.

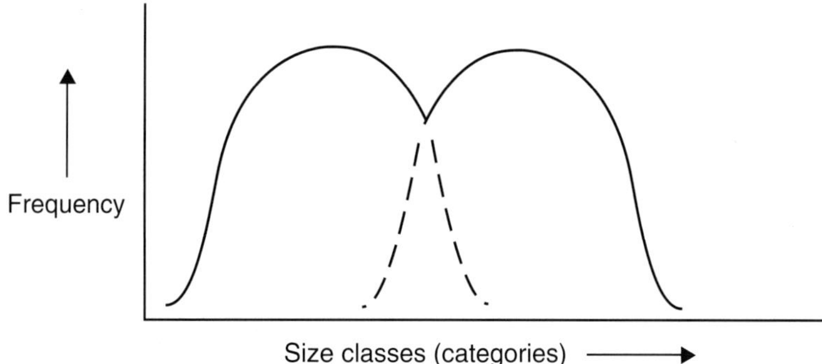

Figure 3.2: Ideal normal curves for two sets of data where the standard deviations are the same but the difference in their means is less obvious

Here the distinction is less clear and so we need to use a statistical method to decide whether there are two distinct populations.

When to use

Note: A minimum population of 30 individuals is required for this test.

This test is useful when you are comparing two populations when each is distributed normally. The test will enable you to discover whether there are in fact two distinct populations. A positive result may act as a starting point for a second hypothesis which aims to determine the reason for two distributions.

Example

An exercise to determine whether the sizes of pebbles deposited on the inside of a river meander are different from those deposited on the outside

Null Hypothesis (H_0)

There is no significant difference between the sizes of pebbles on the inside and on the outside of the meander, i.e. pebbles of all sizes will be deposited randomly across the river bed. (This is the example mentioned on page 10.)

A sample of stones was taken from the inside of a meander and the 'a' axis (longest axis) of each stone was measured and recorded. The procedure was repeated on the outside of the same meander although the sample size was different.

The pebbles appear to form two distinct populations (see Figure 3.3) with the larger ones being distributed on the outside of the meander. However, there is some overlap. Calculating the **standard error of the**

difference in means will allow us to judge whether we can reasonably accept that the pebbles have been differentially deposited.

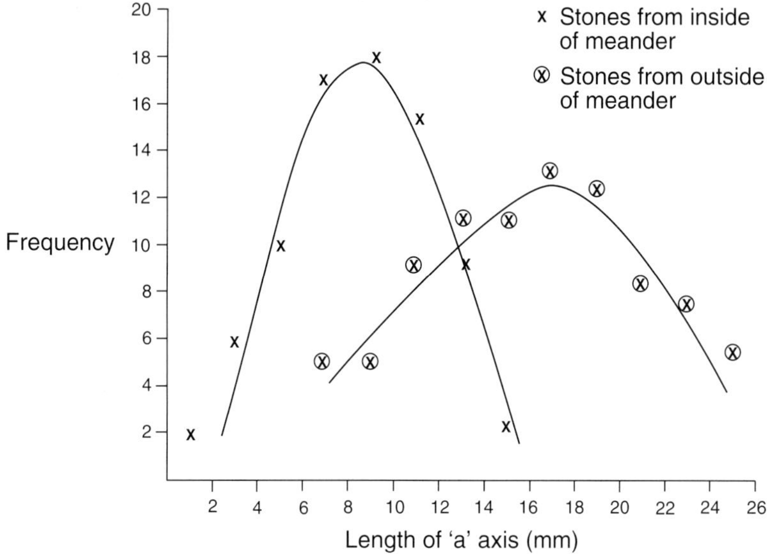

Figure 3.3: Distribution of stone sizes on the inside and on the outside of a meander, using data from Table 3.1 on page 20

The **standard error** of a statistic is the standard deviation of the sampling distribution of that statistic. In this case the sampling distribution of the mean is of interest. If, from our pebble data, we calculated the mean of every five pebble sizes we could then plot all these mean values as in Figure 3.4. These are the sample means.

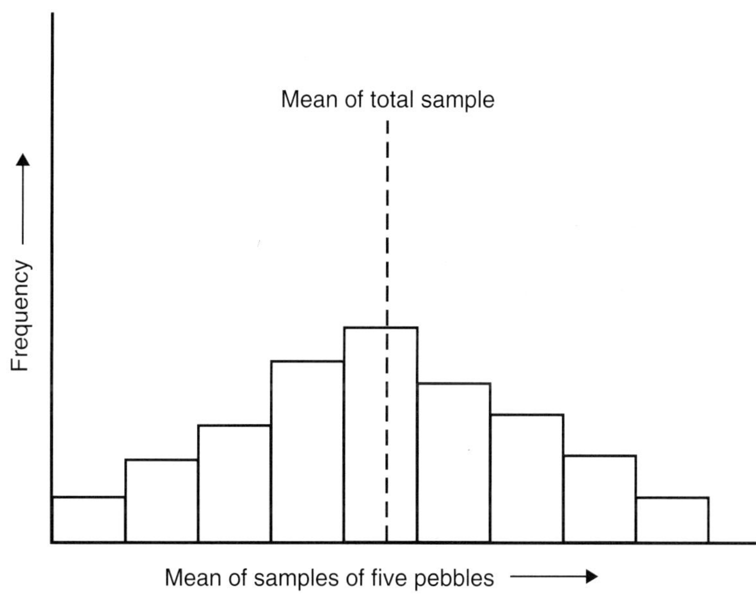

Figure 3.4: Distribution of the means of samples of five pebbles within the whole population

Most of the sample mean values would be expected to be close to the overall (population) mean, with fewer sample means at increasing distances from the population mean — i.e. a normal curve. This curve has a standard deviation which, to distinguish it from the standard deviation of the original measurements, is called the standard error of the mean.

Results

The data is shown in Table 3.1 and analysed in Table 3.2 (both on page 20).

Note that the data has been grouped into size classes. When calculating the means from this type of data, use the midpoint of each size class; for example, the class 0 – 1.9 is regarded as 1.

Returning to our original problem of deciding whether there is a significant difference in stone sizes, the principle is to compare the standard error of the difference in means with the observed difference in means.

'a' axis (mm)	Inside meander (x_1)	Outside meander (x_2)
0– 1.9	2	–
2– 3.9	6	–
4– 5.9	10	–
6– 7.9	17	5
8– 9.9	18	5
10–11.9	15	9
12–13.9	9	11
14–15.9	2	11
16–17.9	–	13
18–19.9	–	12
20–21.9	–	8
22–23.9	–	7
24–25.9	–	5

Table 3.1

	Pebbles inside the meander (x_1)	Pebbles outside the meander (x_2)
Sample size (n)	79	86
Mean ($\bar{x}$)	8.39	16.16
Standard deviation (σ)	3.22	4.88

Table 3.2

$$\text{SE of difference} = \sqrt{\frac{\sigma_1^2}{n_1}+\frac{\sigma_2^2}{n_2}}$$

$$= \sqrt{\frac{10.37}{79}+\frac{23.81}{86}}$$

$$= \sqrt{0.13+0.28}$$

$$= \mathbf{0.64}$$

Equation 3.1

The standard error of the difference in means is calculated using Equation 3.1 (left).

A difference between two values is judged to have attained statistical significance if it equals or exceeds twice the standard error. The probability of exceeding twice the SE by chance is approximately one-twentieth or 5%. If our observed difference between the means equals or exceeds 2 ˘ SE then we are confident that there is a 95% probability that the two populations are distinct.

Observed difference of means = 16.16 – 8.39 = 7.77

Standard error of difference of means = 0.64

Clearly the observed difference (7.77) easily exceeds twice the standard error of the difference (1.28). Therefore the null hypothesis may be rejected, and we conclude that there is a significant difference between the stone sizes on the inside and the outside of this river meander.

It may be tempting to ascribe a reason for this observed difference, i.e. the effect of the different river velocities sorting the stones so that in the faster current the smaller stones are removed. Whilst this is very likely, it must be emphasised that this data does not allow us to draw this conclusion. To test the new hypothesis a different type of data must be collected.

Significance and confidence limits

The standard deviation may be used to show whether two sets of data do differ significantly. The term **significant** in a statistical sense has a precise mathematical definition. Significance concerns the reliability of the data and is expressed as a percentage value.

For example, if we say that the information is significant at the 95% level, then this means that only 5 times out of every 100 would this data occur by chance. Similarly, 99% significance means that this data would only occur once every hundredth time purely by chance.

The degree of significance required depends upon the nature and application of the data. For field data the 95% level is usually appropriate. In drugs trials a much greater degree of reliability is required in view of the possible unfortunate consequences of using the product. In this case only a 99.9% significance may be acceptable.

In significance tables the 5% and 1% levels are indicated as 0.05 and 0.01 respectively. If the calculated value exceeds the theoretical value in either (or both) columns, then that value is significant at the appropriate level. We could say that we are **confident** at the 95% or 99% level of the reliability of our data. Hence the term 'confidence limits' refers to the degree of reliability that we require.

The 't' distribution

When to use

Note: Never use the t-test when there are more than thirty individuals in a sample.

One of the major limitations of the comparison of two means is that a minimum of thirty individuals within each population is required. **Sometimes in fieldwork it is not possible to collect that amount of data due to limitations of numbers of available individuals or of time. In those cases it is still possible to compare two sample populations using the Student's t-test.**

The t-test uses the standard error of the differences between means (as in the previous test). However the 't' distribution is used rather than the normal distribution. The 't' distribution is symmetrical about a mean of zero and is bell-shaped (as is the normal distribution) but is flatter, more dispersed. The fewer the individuals within the sample, the flatter is the curve (Figure 3.5).

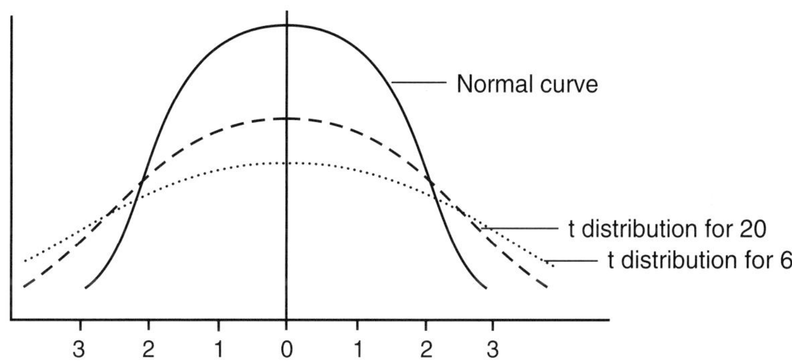

Figure 3.5: The 't' distribution and the normal curve

As the sample size tends towards 30 then the 't' distribution tends towards the normal distribution, and hence the former test (the comparison of two means) can be used.

The formula for calculation of the t-test is:

$$t = \frac{|\bar{x} - \bar{y}|^{*}}{\sqrt{\frac{(\Sigma x^2/n_x) - \bar{x}^2}{n_x - 1} + \frac{(\Sigma y^2/n_y) - \bar{y}^2}{n_y - 1}}}$$

Equation 3.2

* $|\bar{x} - \bar{y}|$ means that the absolute difference in sample means is taken, i.e. the value must be positive.

where $\bar{x}$ and $\bar{y}$ are each sample mean,
Σx and Σy are the sums of each sample,
n_x and n_y are the number of individuals in each sample.

Example

The following problem illustrates the use of the t-test. Is there a significant difference between the number of plant species supported by acid moorland and limestone upland?

Null Hypothesis (H_0)
Acid moorland supports the same number of plant species as limestone upland.

A series of quadrats were thrown in each habitat and the total number of species in each quadrat was recorded.

Number of species

The results are given in Table 3.3 below, and the calculation of t is shown below left (Equation 3.3).

Table 3.3

Quadrat No.	Acid (x)	Limestone (y)	x^2	y^2
1	6	14	36	196
2	8	12	64	144
3	9	6	81	36
4	4	11	16	121
5	7	15	49	225
6	11	14	121	196
7	7	17	49	289
8	6	8	36	64
9	8		64	
10	7		49	
	$\Sigma x=73$ $\bar{x}=7.3$ $n_x=10$	$\Sigma y=97$ $\bar{y}=12.13$ $n_y=8$	$\Sigma x^2=565$	$\Sigma y^2=1271$

$$t = \frac{|\bar{x}-\bar{y}|}{\sqrt{\frac{(\Sigma x^2/n_x)-\bar{x}^2}{n_x-1}+\frac{(\Sigma y^2/n_y)-\bar{y}^2}{n_y-1}}}$$

$$= \frac{4.83}{\sqrt{\frac{(565/10)-53.29}{10-1}+\frac{(1271/8)-147.13}{8-1}}}$$

$$= \frac{4.83}{\sqrt{\frac{3.21}{9}+\frac{11.35}{7}}}$$

$$= \frac{4.83}{1.43} = \mathbf{3.38}$$

Equation 3.3

The degrees of freedom (see Glossary) are given by $(n_x - 1) + (n_y - 1) = 16$. Consulting tables of significance (Appendix 2), 3.38 is greater than the critical value of t at $p = 0.05$ and hence the null hypothesis may be rejected. Consequently we can assert the statement that there is a significant difference in the number of plant species supported by acid moorland and limestone upland.

The comparison of two medians — The Mann-Whitney U-test

When to use

Note: This test should be used when both sample sizes are below 20 and above 5.

As in the comparison of two means and the t-test, this test will determine whether the data obtained belongs to two separate populations. However, the important difference is that this test determines whether there is a significant difference between the two medians rather than the means.

The advantage of this test is that there is no assumption that the data is normally distributed. However, both samples must exhibit similarly shaped distributions. If in doubt, plot them first. It should be used when both sample sizes are below 20 and above 5. The null hypothesis is of the form that there is no significant difference between the two samples, i.e. that all the data is obtained from the same population.

Null Hypothesis (H_0)

There is no significant difference in the breaking strains of cotton recovered from peat and podsol. (This is the example already mentioned on pages 11 and 12.)

Example

The experiment described in Chapter 1 was designed to determine whether there was any significant difference in the rates of decomposition in the peat and podsol soils. Ten strands of cotton were buried in each site and recovered some weeks later. The breaking strain of each strand was tested and recorded. Any reduction in original breaking strain was regarded as due to decomposition of the cellulose strands.

Results

The results are set out as in Table 3.4 (page 23) and each result is ranked in order (lowest value first). Note that the total sample is ranked together and not the individual columns. The sample size in each case is recorded (they may be different) and each rank order column is summed and recorded.

Peat		Podsol	
Reduction in breaking strain (%)	Rank	Reduction in breaking strain (%)	Rank
x	r_x	y	r_y
23	5	29	9
14	1	42	17
19	3	52	20
33	11	* 37	13.5
24	6	46	18
17	2	49	19
25	7	28	8
21	4	31	10
* 37	13.5	36	12
39	16	38	15
No. in sample (N_x) = 10		No. in sample (N_y) = 10	
Total of rank scores (Σr_x) = 68.5		Total of rank scores (Σr_y) = 141.5	

Table 3.4

* In the case of two (or more) equal results their mean rank is expressed in both cases.

The Mann-Whitney U-statistic is calculated using Equations 3.4 and 3.5:

$$U_x = N_x \cdot N_y + \frac{N_x \cdot (N_x + 1)}{2} - \Sigma r_x$$

$$= 10 \times 10 + \frac{10 \times 11}{2} - 68.5$$

Equation 3.4

$$= \mathbf{86.5}$$

$$U_y = N_x \cdot N_y + \frac{N_y \cdot (N_y + 1)}{2} - \Sigma r_y$$

$$= 10 \times 10 + \frac{10 \times 11}{2} - 141.5$$

Equation 3.5

$$= \mathbf{13.5}$$

Note that $U_x + U_y$ must equal $N_x \cdot N_y$ (if they do not, then recheck the arithmetic). The value of U to be used in determining significance is always the smaller of the two, in this case U_y. Consult the significance table in Appendix 3. If the value of U is equal to or smaller than the relevant value in the table, then the null hypothesis may be rejected.

At $p = 0.05$ where $N_x = 10$, $N_y = 10$ the stated value is 23. Clearly we can reject the null hypothesis. Therefore there is a significant difference in the median rates of decomposition of cellulose in a peat and a podsol.

Finally, remember that this is a test of difference between medians and not of difference between means. Thus you must express your conclusion in those terms.

4: THE CHI-SQUARED (χ^2) TEST

When to use

The Chi-squared (χ^2) test is used when the object of the exercise is to attempt to observe differences between comparable sets of data. It is particularly useful when the expected distribution of the data is not known.

The data must be collected in such a way that it is capable of being grouped into classes. In each case a null hypothesis must be put forward. This is usually of the form:

'Z' is distributed at random.

The χ^2 value, determined by the appropriate formula, may then be compared with significance tables. These will confirm whether any deviation from random in the observed data is merely a chance effect or does possess statistical significance.

Note: If any expected value is below 5 then the test becomes invalid.

If statistical significance is achieved then the null hypothesis may be rejected (5% is considered a satisfactory level from fieldwork data). Please note, however, that even though you may have reliably rejected a random distribution, you have not proved any causal effect. To do so could be a logical extension of a project, having achieved a statistical relationship.

The main limitation of the test is when only small samples are available. If any *expected value* is below 5 then the test becomes invalid.

Example 1

Aim

To identify any preferred orientation of corries in Snowdonia.

Methods

Corries were identified from the relevant OS maps and their individual orientations were recorded. These were grouped in four categories relating to the four quadrants of the compass.

Orientation from true north	Frequency of corries
0°– 89°	30
90°–179°	5
180°–269°	6
270°–359°	11
	Total = **52**

Table 4.1

The results are shown in Table 4.1.

Whether the observed distribution of corrie orientation is significant or merely a chance effort can be tested by the χ^2 test.

Null Hypothesis (H_0)

The orientation of corries is random, i.e. there is no significant alignment.

If our null hypothesis (left) is correct, then we would expect there to be 52 ÷ 4 = 13 corries in each category. This is obviously not the case (from comparison with Table 4.1) and the χ^2 test will establish whether the departure from perfectly random distribution is a chance effect or statistically significant.

The χ^2 formula is:

$$\chi^2 = \sum \frac{(O - E)^2}{E}$$

Equation 4.1

where Σ = the sum of
O = observed frequency
E = expected frequency.

Our corrie data may be set out as in Table 4.2 (page 25). Note that no observed or expected value falls below 5.

From Table 4.2 we see that $\chi^2 = 31.23$. This figure has to be checked with significance tables for χ^2. It is necessary to calculate **degrees of freedom** (see Glossary) which equals (n – 1) where n = the number of categories.

Hence in this case there are 3 degrees of freedom.

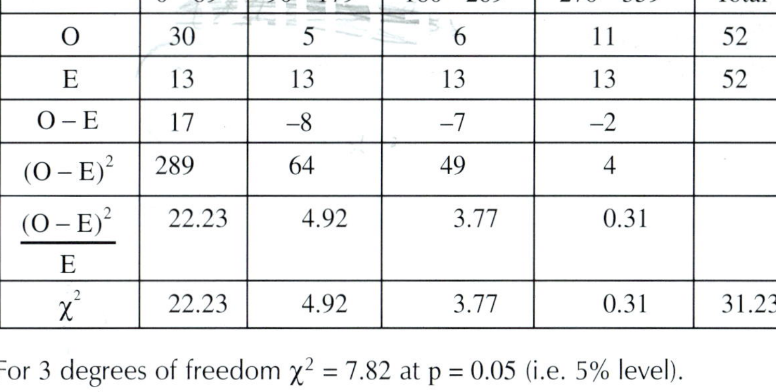

	0°–89°	90°–179°	180°–269°	270°–359°	Total
O	30	5	6	11	52
E	13	13	13	13	52
O – E	17	–8	–7	–2	
$(O-E)^2$	289	64	49	4	
$\frac{(O-E)^2}{E}$	22.23	4.92	3.77	0.31	
χ^2	22.23	4.92	3.77	0.31	31.23

Table 4.2

For 3 degrees of freedom $\chi^2 = 7.82$ at $p = 0.05$ (i.e. 5% level).

Hence the null hypothesis can be rejected and we can confirm that there is some preferred orientation.

Note: It is not valid to state that we have established the reason for this, merely that we are statistically satisfied that the corries are not randomly aligned.

Example 2

It is common when collecting results to find that two variables being studied can be each divided into two categories. A more sophisticated contingency table (see page 26) can be drawn up.

Aim

To detect any differences in snail-shell patterns in two distinct habitats.

Methods

It is well known that certain snails may exhibit different shell banding patterns within the species. The purpose of this experiment was to see whether lighter-coloured snails predominated on bare limestone pavement whereas darker snails were more successful in an adjacent limestone woodland. Large samples of snails were taken in each habitat and later classified according to shell colour.

Table 4.3: Number of light- and dark-coloured individuals of the snail *Cepaea hortensis* found in two distinct habitats

	Light shells	Dark shells
Limestone pavement	153	112
Limestone woodland	96	120

> ***Null Hypothesis (H_0)***
> There is no significant difference in snail-shell colour in either habitat.

The χ^2 contingency table should be set out as in Table 4.4:

Table 4.4

	Light shells	Dark shells	Total
Limestone pavement	a	b	a + b
Limestone woodland	c	d	c + d
Total	a + c	b + d	n

Equation 4.2

$$\chi^2 = \frac{n(ad-bc)^2}{(a+b)(c+d)(a+c)(b+d)}$$

Therefore for our example:

Table 4.5

	Light shells	Dark shells	Total
Limestone pavement	153	112	265
Limestone woodland	96	120	216
Total	249	232	481

$$\chi^2 = \frac{481 \times (153 \times 120 - 112 \times 96)^2}{265 \times 216 \times 249 \times 232}$$

Equation 4.3

$$= \mathbf{8.42}$$

In a contingency table there are $(r - 1)(c - 1)$ degrees of freedom, where r = number of rows and c = number of columns.

Hence in this case there is 1 degree of freedom.

Again consulting χ^2 tables, the result is significant even at $p = 0.01$ and therefore the null hypothesis may be rejected.

Again note that no reasons for this difference are established by the test.

More complex situations

The Chi-squared test can be developed further so that more complex situations may be analysed. Example 2 concerns the situation where two variables may be distributed amongst two classes. It is perfectly feasible to extend a 2 ˘ 2 contingency table to h ˘ k, where h and k represent any number.

Care must be taken to avoid a value of less than 5 in any cell in the table. If expected numbers must fall below 5 then an alternative formula with the addition of Yates' correction may be used. This is:

Equation 4.4

$$\chi^2 = \frac{(|ad - bc| - \tfrac{1}{2}n)^2}{(a+b)(c+d)(a+c)(b+d)}$$

The vertical lines either side of $|ad - bc|$ indicate that the positive value must be taken.

Summary of procedure with expected ratio

1 Formula for χ^2:

$$\chi^2 = \sum \frac{(O - E)^2}{E}$$

2 Establish null hypothesis

3 Calculate expected values

4 Degrees of freedom = $n - 1$

Summary of procedure with no expected ratio

1 Establish null hypothesis

2 Construct contingency table. Check that no cell has a value less than 5; preferably each cell should be above 10.

	X_1	X_2	Total
Y_1	a	b	a + b
Y_2	c	d	c + d
Total	a + c	b + d	n

3 Apply the formula:

$$\chi^2 = \frac{n(ad - bc)^2}{(a+b)(c+d)(a+c)(b+d)}$$

4 Degrees of freedom = $(r - 1)(c - 1)$

5 If any expected values fall below 5 then use Yates' correction:

$$\chi^2 = \frac{(|ad - bc| - \tfrac{1}{2}n)^2}{(a+b)(c+d)(a+c)(b+d)}$$

Remember: The vertical lines surrounding $|ad - bc|$ indicate that the positive value must be taken.

5: CORRELATIONS

Often a fieldwork exercise aims to relate factors and show that each varies either positively or negatively with the other. For example, it can be shown that the wind-speed increases and temperatures drop as we climb uphill. Clearly there is some relationship here, but how strong is it? **A correlation coefficient will measure the strength of any observed relationship.**

The simplest way to illustrate a relationship between two variables is first to construct a scattergraph and then to draw a 'line of best fit' through these points. The resulting graph may look like one of the examples in Figure 5.1.

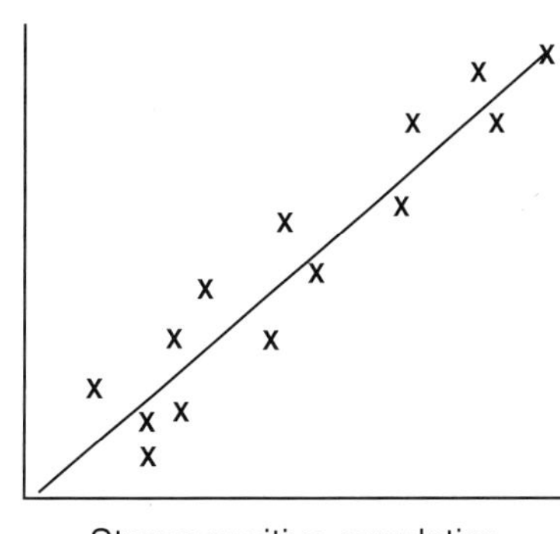

Strong positive correlation

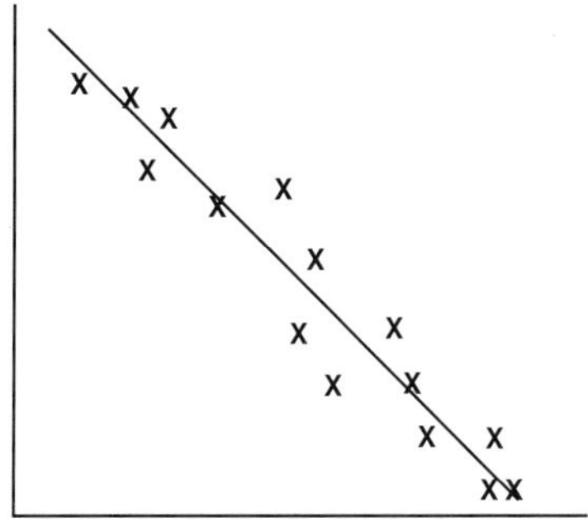

Strong negative correlation

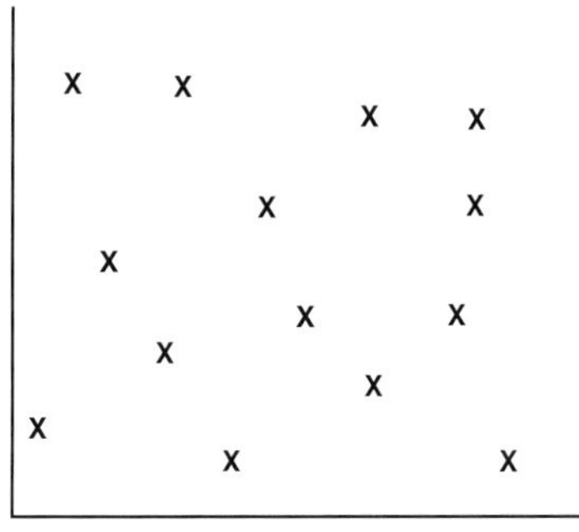

No correlation

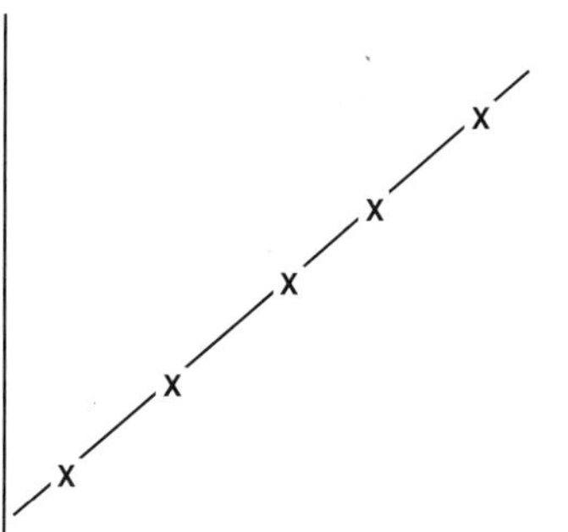

Perfect positive correlation

Figure 5.1: Various forms of scattergraphs

The scattergraph will illustrate possible relationships but it will not give an indication of the strength of the relationship. For this we must use a **correlation coefficient** (r). The value of r lies between –1 and +1. The minus sign indicates a negative relationship, and the nearer to –1 the stronger the negative correlation. Similarly, positive values indicate a positive relationship, and once again the nearer to +1 the stronger the positive correlation.

A positive relationship means that as variable X increases so does variable Y. A negative relationship indicates that as one variable increases the other decreases.

Two correlation coefficients will be described here: the Spearman Rank Correlation Coefficient, which uses data according to its rank order, and the Product-Moment Correlation Coefficient, which uses the actual values of the data.

Spearman Rank Correlation Coefficient

This test makes use of ranked ordinal data and is particularly useful as a good general assessment of a relationship. Because it does not use absolute values of the data, its accuracy is somewhat less than that of the Pearson Product-Moment test (see page 30).

Null Hypothesis (H_0)

Altitude has no effect on seed production of the rush.

Example 1

Measurements were taken of seed production in the rush *Juncus effusus* on moorland sites at different altitudes. At each altitude a large number of seedheads were collected and the mean number of seeds per seedhead was calculated. The data is shown in Table 5.1 (below left).

Altitude (ft)	Mean no. of seeds
700	126.5
850	94.4
950	103.4
1050	50.5
1100	65.2
1150	64.1
1250	75.8
1394	52.8
1450	63
1500	52.4

Table 5.1

Figure 5.2 (below) shows the data plotted as a scattergraph, suggesting a negative correlation.

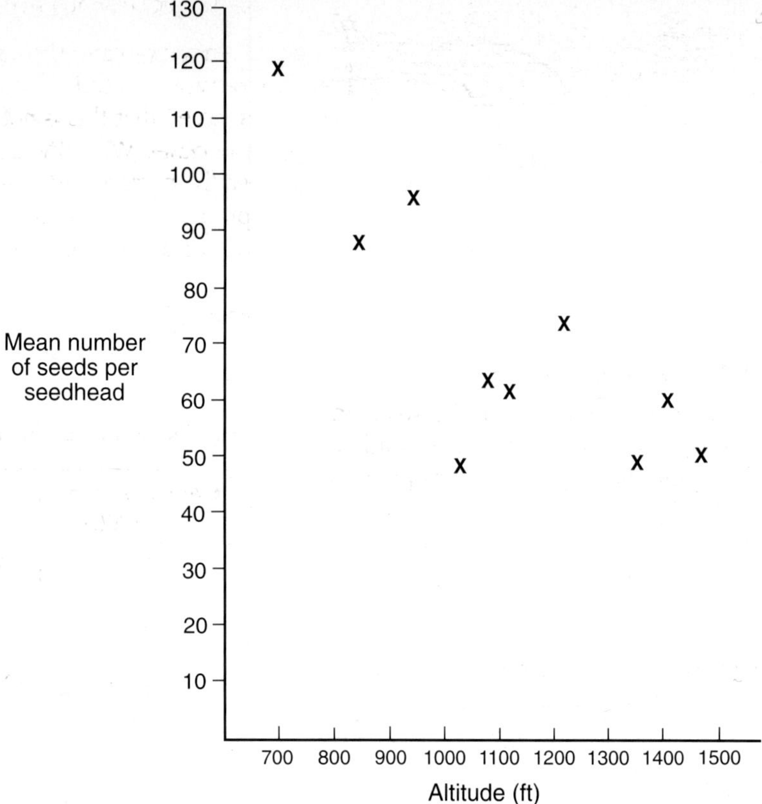

Figure 5.2: Relationship between altitude and seed production of the soft rush *(Juncus effusus)*

The information should be set out as in Table 5.2 below.

Variable A Altitude in feet	Rank	**Variable B** Seeds (mean no. per head)	Rank	d	d^2
700	10	126.5	1	9	81
850	9	94.4	3	6	36
950	8	103.4	2	6	36
1050	7	50.5	10	–3	9
1100	6	65.2	5	1	1
1150	5	64.1	6	–1	1
1250	4	75.8	4	0	0
1394	3	52.3	9	–6	36
1450	2	63	7	–5	25
1500	1	52.4	8	–7	49
				$\Sigma d^2 = 274$	

Table 5.2

The items are put into rank order where the highest value ranks as 1. d is the difference between each pair of rank variables.

The formula for Spearman Rank ρ (rho) is shown in Equation 5.1:

$$\rho = 1 - \left(\frac{6\Sigma d^2}{n^3 - n} \right)$$

Equation 5.1

where n is the number of observed pairs (in this case 10). There are n (= 10) degrees of freedom.

$$\rho = 1 - \left(\frac{6 \times 274}{1000 - 10} \right)$$

$$= 1 - \left(\frac{1644}{990} \right) = -\mathbf{0.66}$$

Equation 5.2

Referring to the appropriate significance tables (Appendix 5) we see that –0.66 lies between the confidence levels 95% and 99%, and so we can reject our null hypothesis.

Hence we have shown that there is a significant negative correlation between altitude and seed production in the soft rush. It must be stressed that this is not necessarily the direct effect of one variable on the other. With altitude, temperatures decrease and wind-speeds generally increase and these may well be the direct influences on the seed production of the soft rush. **Detection of a significant correlation does not imply a causal relationship.**

Example 2

The relationship between car ownership and population density in a British city.

Null Hypothesis (H_0)
There is no relationship between the proportion of cars per head of population and the population density.

The data are given in Table 5.3 below:

Car ownership (cars/1000)	Rank	Population per unit area	Rank	d	d^2
126	7	10.1	8	–1	1
151	2	8.4	10	–8	64
101	11	22.0	2	9	81
106	10	17.3	3	7	49
132	4	9.1	9	–5	25
79	12	27.3	1	11	121
145	3	16.5	4	–1	1
107	9	15.0	5	4	16
127 *	5.5	3.5	12	–6.5	42.25
182	1	11.3	7	–6	36
118	8	11.5	6	2	4
127 *	5.5	4.4	11	–5.5	30.25
				$\Sigma d^2 = 470.5$	

* When two values are identical then the average rank value is given to each.

Table 5.3

$$\rho = 1 - \left(\frac{6\Sigma d^2}{n^3 - n} \right)$$

$$\rho = 1 - \left(\frac{6 \times 470.5}{1728 - 12} \right)$$

$$= 1 - \left(\frac{2823}{1716} \right) = -\mathbf{0.65}$$

Equation 5.3

For 12 degrees of freedom this result is significant between 95% and 99%. The null hypothesis can therefore be rejected. Hence we can state that there is a significant negative relationship between car ownership and population density.

As before, care must be taken in interpreting this result. We cannot state categorically that because people have more space to live in then their need (or ability) to possess more cars is increased. More information must now be collected concerning the nature of the areas studied and the socio-economic rating of each of the populations.

Spearman Rank testing is useful and reasonably quick to calculate. The main limitation is its inherent inaccuracy due to the ranking of data. Because of this the magnitude of the difference in values is ignored. The Pearson Product-Moment test uses actual values and thus overcomes this problem. However, the main advantage of the Spearman Rank test over the Pearson Product-Moment test is that it does not assume that each variable is distributed normally.

Note: Rank largest value as 1;
d = difference between pairs of ranked observations

Spearman Rank Field Recording Sheet

Site name = ______________________

Name and units of variable A = ______________________

Name and units of variable B = ______________________

Field data variable A	Rank	Field data variable B	Rank	d	d^2
				$\Sigma d^2 =$	

$$\rho = 1 - \left(\frac{6\Sigma d^2}{n^3 - n}\right) = \_\_\_\_\_\_\_\_\_\_$$

The Pearson Product-Moment Correlation Coefficient

This is a more sophisticated test than the Spearman Rank test. It gives a more accurate result because it uses the actual measured values of the data rather than their relative rankings. For the test to be used with validity, however, the data must come from a normally distributed population. If unsure then use the Spearman Rank Correlation test.

Example

Aim

To find the relationship between discharge of a northern hill stream and distance downstream.

Null Hypothesis (H_0)

There is no relationship between the discharge of a river and distance from its source.

Ten sites were chosen along the river Hodder, and at each point the discharge of the river was calculated using measurements of cross-sectional area and mean velocity.

Results

The results are shown in Table 5.4 (below).

Distance from source (m)	Discharge (cumecs)
1 800	0.25
2 400	0.39
3 400	0.68
4 200	0.42
8 000	0.46
12 000	1.30
13 000	1.52
13 600	1.69
14 000	1.85
15 000	2.75

Table 5.4

There does appear to be a relationship which, if suspected, is best illustrated by drawing a scattergraph (Figure 5.3).

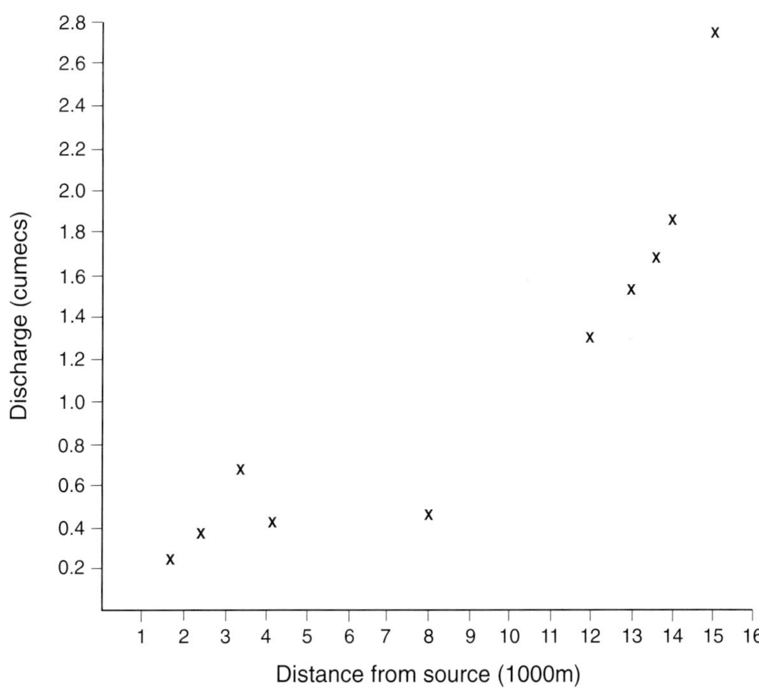

Figure 5.3: Relationship between the distance from the source of the river Hodder and its discharge

The scattergraph suggests a positive relationship, and the correlation test is used to discover its strength. The Pearson Product-Moment test is

Distance downstream (m) (x)	Discharge (cumecs) (y)	Deviation of x from $\bar{x}$ (dx)	Deviation of y from $\bar{y}$ (dy)	d^2x	d^2y	dx.dy
1 800	0.25	6 940	0.88	48 163 600	0.7744	6 107.2
2 400	0.39	6 340	0.74	40 195 600	0.5476	4 691.6
3 400	0.68	5 340	0.45	28 515 600	0.2025	2 403
4 200	0.42	4 540	0.71	20 611 600	0.5041	3 223.4
8 000	0.46	740	0.67	547 600	0.4489	495.8
12 000	1.30	–3 260	–0.17	10 627 600	0.0289	554.2
13 000	1.52	–4 260	–0.39	18 147 600	0.1521	1 661.4
13 600	1.69	–4 860	–0.56	23 619 600	0.3136	2 721.6
14 000	1.85	–5 260	–0.72	27 667 600	0.5184	3 787.2
15 000	2.75	–6 260	–1.62	39 187 600	2.6244	10 141.2
Σx=87 400 $\bar{x}$=8 740	Σy=11.31 $\bar{y}$=1.13			Σd^2x= 2.57 × 10^8	Σd^2y= 6.1149	Σ(dx.dy)= 35 786.6

Table 5.5

a lengthy task without the use of a calculator, but if the data is set out using a structured table as illustrated in Table 5.5 then few problems should arise. This will also help those who wish to design an appropriate spreadsheet — see section on Computer Software (page 63).

The Pearson Product-Moment Correlation Coefficient (r) is given by a formula (Equation 5.4) which may appear somewhat intimidating:

Equation 5.4

$$r = \frac{\Sigma(dx \cdot dy)}{\sqrt{\Sigma d^2x \cdot \Sigma d^2y}}$$

$$r = \frac{35786.6}{\sqrt{2.57 \times 10^8 \times 6.1149}}$$

Equation 5.5

$$= \frac{35\,786.6}{39\,648} = \mathbf{0.903}$$

A positive value for r indicates a positive correlation.

To test the significance of this result we consult the table (Appendix 6). There are N – 1 degrees of freedom where N is the number of paired observations. In this case there are 9 degrees of freedom. To reject our null hypothesis we require 95% confidence.

At significance level p = 0.05 and for 9 degrees of freedom r must exceed 0.521. In fact this result can be accepted at significance level p = 0.01 and hence the null hypothesis may be rejected. Therefore there is a positive relationship between discharge and distance from the source of the river Hodder.

Cautionary points about correlation coefficients

1 Correlation is a useful tool for identifying certain relationships between two variables. Choice between the Spearman Rank and the Pearson Product-Moment methods depends upon the degree of accuracy required and also whether the variables are from a normal distribution. Clearly the Spearman Rank method is quicker and gives a fairly accurate result which is usually sufficient within the precision limits of fieldwork data. The Spearman Rank method, unlike the Pearson Product-Moment method, does not assume that the variables measured are distributed normally. **A very precise statistical analysis of unrigorous data will not yield results which are any more reliable.** In other words:

garbage in = garbage out!

There are, however, cases on the borderlines of significance which may be rejected using the Spearman Rank coefficient which are acceptable when tested using the Pearson Product-Moment coefficient.

2 A significant result does not imply any underlying reasons for the association. **Neither does it imply that one variable is the cause of variation in the second.** The fact that two plants are often found together, for example, does not necessarily mean that species X depends upon the presence of species Y. It may well be that both species have very similar habitat requirements.

3 **Nonsense correlations are possible.** It has been shown that the number of storks sighted per year correlates positively and significantly with the number of births in Sweden over a particular decade! Only perform correlation analysis where there is a possible relationship between the two variables.

4 **Correlation is a measure of linear relationships only.** Before you embark on the calculations comparing two sets of information, it is essential to draw a scattergraph to ensure that the relationship is linear (see Figure 5.4).

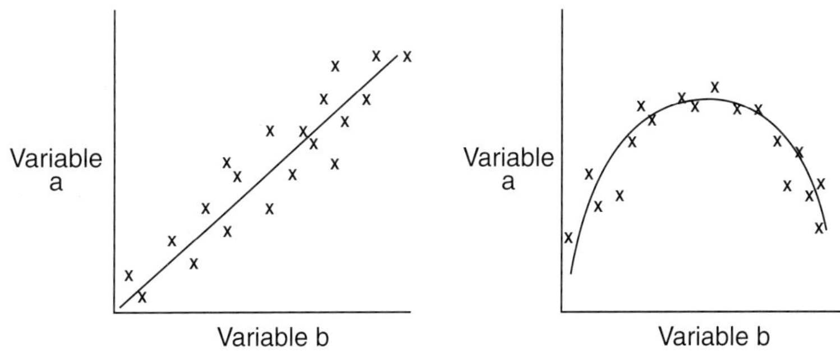

Figure 5.4: Linear and non-linear relationships between variables

6: LINEAR REGRESSION

When to use

Unless the points all lie in a straight line, then a 'line of best fit' cannot be plotted by eye with any degree of certainty (see Figure 6.1).

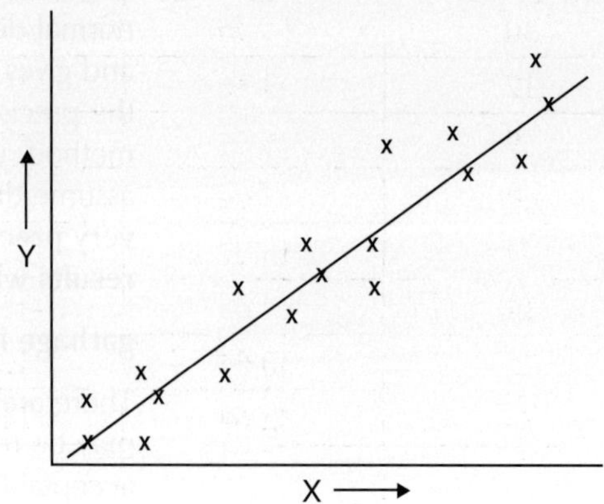

Figure 6.1: 'Line of best fit' when plotted by eye

There is an apparent positive relationship between variables X and Y. Calculation of the correlation coefficient would confirm this. A 'line of best fit' may be plotted by eye and this would be useful if you wished to predict the value of (say) Y for a known value of X. Clearly this is a subjective method and is prone to error.

The 'line of best fit' can be calculated mathematically using the regression equation (see Equations 6.2 and 6.3 on page 35).

Example

Soil samples were taken from various sites from within a sand dune succession. Each was analysed back in the laboratory, and the results are shown in Table 6.1 (left). There was a possible relationship between the water content and organic content of each sample.

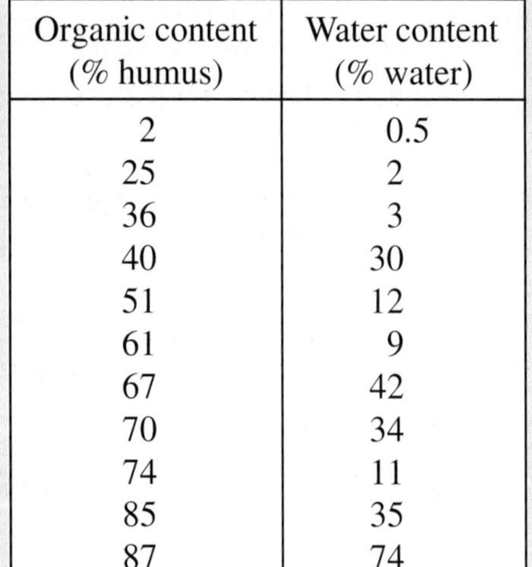

Organic content (% humus)	Water content (% water)
2	0.5
25	2
36	3
40	30
51	12
61	9
67	42
70	34
74	11
85	35
87	74

Table 6.1

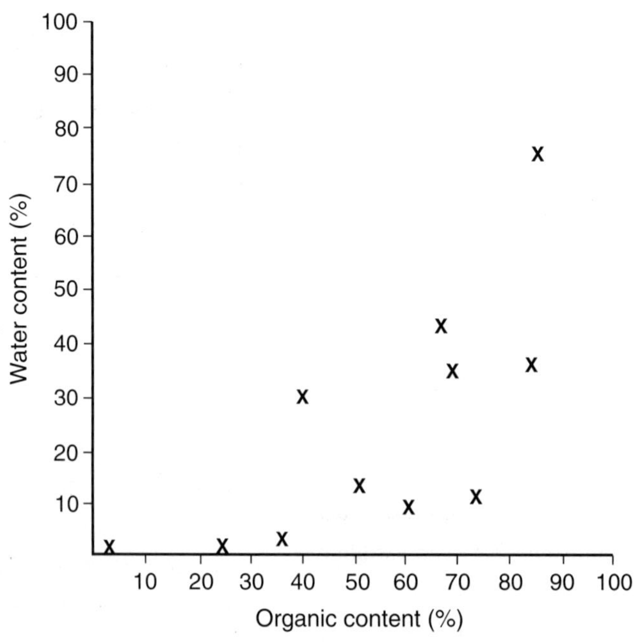

Figure 6.2: Relationship between soil water content and soil organic content

The scattergraph does suggest a positive relationship between the two variables which can be tested using the Pearson Product-Moment Correlation Coefficient (see previous chapter).

Null Hypothesis (H_0)

There is no relationship between the organic content and the water content of the soils.

Organic content (%) (x)	Soil water (%) (y)	Deviation of x from $\bar{x}$ (dx)	Deviation of y from $\bar{y}$ (dy)	d^2x	d^2y	dx·dy*
2	0.5	–52.36	–22.45	2 741.57	504.00	+1 175.48
25	2	–29.36	–20.95	862.01	438.90	+615.09
36	3	–18.36	–19.95	337.09	398.00	+366.28
40	30	–14.36	7.05	206.21	49.70	–101.24
51	12	–3.36	–10.95	11.29	119.90	+36.79
61	9	6.64	–13.95	44.09	194.60	–92.63
67	42	12.64	19.05	159.77	362.90	+240.79
70	34	15.64	11.05	244.61	122.10	+172.82
74	11	19.64	–11.95	385.73	142.80	–234.70
85	35	30.64	12.05	938.81	145.20	+369.21
87	74	32.64	51.05	1 065.37	2 606.10	+1 666.27
$\Sigma x = 598$ $\bar{x} = 54.36$	$\Sigma y = 252.5$ $\bar{y} = 22.95$			$\Sigma d^2x =$ 6 996.55	$\Sigma d^2y =$ 5 084.2	$\Sigma(dx.dy) =$ 4 214.16

Table 6.2

* In Table 6.2, take care with the plus and minus signs when multiplying dx by dy.

The Pearson Product-Moment Correlation Coefficient is calculated as previously described.

$$r = \frac{\Sigma(dx \cdot dy)}{\sqrt{\Sigma d^2x \cdot \Sigma d^2y}}$$

$$r = \frac{4214.16}{\sqrt{6996.55 \times 5084.2}}$$

Equation 6.1

$$= \mathbf{0.71}$$

This is significant at a level greater than 0.01. Hence there is a relationship between soil organic content and soil water content.

The data is quite scattered, and as we wish to plot a line of best fit with confidence the regression line must be calculated.

Calculating the regression line

The regression line can be calculated using the following equation. For a line y on x (that is how y alters when x is altered) y = a + bx where:

a is the point taken by y when x is zero (this is called the intercept);

b is the regression coefficient which describes the gradient of the line.

To calculate b, use the equation:

$$b = \frac{\Sigma(dx \cdot dy)}{\Sigma d^2x}$$

Equation 6.2

$$b = \frac{4214.16}{6996.55} = \mathbf{0.60}$$

To find the exact intercept point, substitute the mean values of x and y (i.e. $\bar{x}$ and $\bar{y}$) into the regression equation:

$$\bar{y} = a + b\bar{x}$$
$$22.95 = a + (0.60 \times 54.36)$$
$$a = 22.95 - 32.62$$

Equation 6.3

$$= \mathbf{-9.67}$$

To plot the regression line we simply calculate values of y for chosen values of x, again using y = a + bx. (Choose one value of x near to the far end of the x axis and one other in the middle range.)

When x = 100, y = –9.67 + (0.60 ˘ 100) = **50.33**

When x = 40, y = –9.67 + (0.60 ˘ 40) = **14.33**

We now have three points — (0, –9.67), (100, 50.33) and (40, 14.33) — which can be plotted onto the scattergraph. They should all lie on the same straight line: if not, check your calculations!

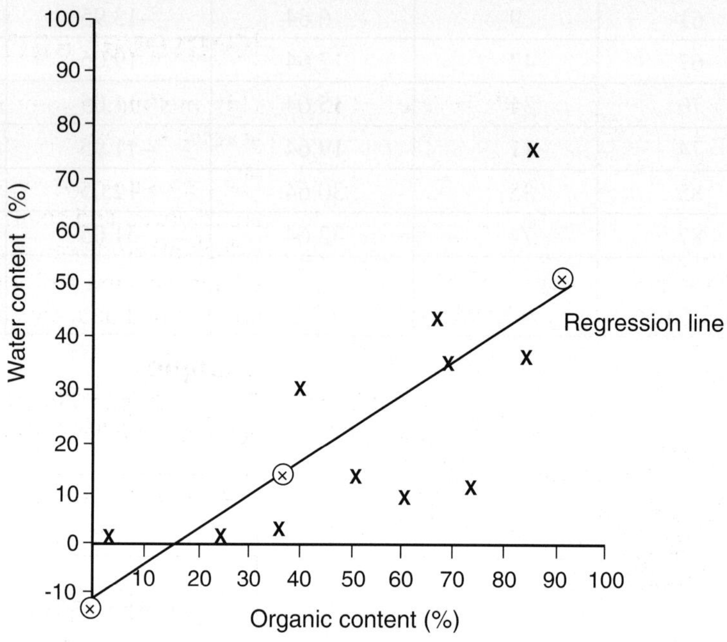

Figure 6.3: Regression line plotted onto Figure 6.2

A regression line, once plotted, allows us to predict values of y for any chosen value of x. It is possible to calculate confidence limits for the regression line, although this is outside the scope of this book.

7: ESTIMATING THE SIZE OF ANIMAL POPULATIONS

Animal populations are often more difficult to measure than plant populations because of their mobility. The following methods may be used where estimates are required.

Removal sampling method

When to use

This method is appropriate where a relatively small and easily accessible area is to be sampled, e.g. a grassland.

The area is thoroughly sampled and all individuals of the chosen species are removed and counted. The area is sampled again and all the individuals of the chosen species are removed a second time. This is repeated until very few individuals are caught. Note that captured animals are not returned until the conclusion of the survey.

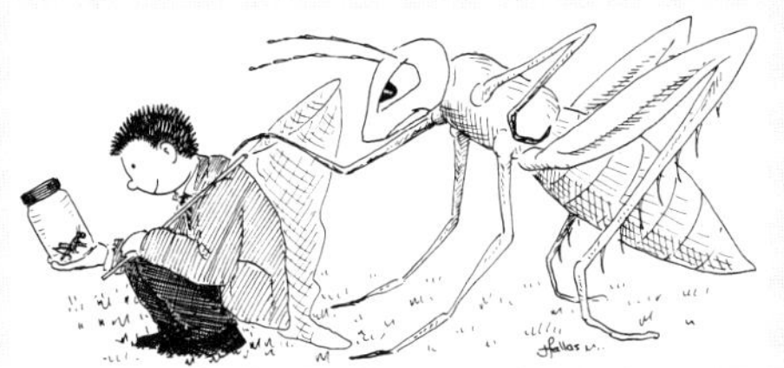

Example

The grasshoppers in an area of grassland were sampled with a sweep net and the following results were obtained:

Table 7.1

Sample number	Number captured	Total number caught in previous samples
1	32	0
2	28	32
3	16	60
4	11	76
5	4	87

Plotting the number captured against total number previously captured gives the graph shown in Figure 7.1.

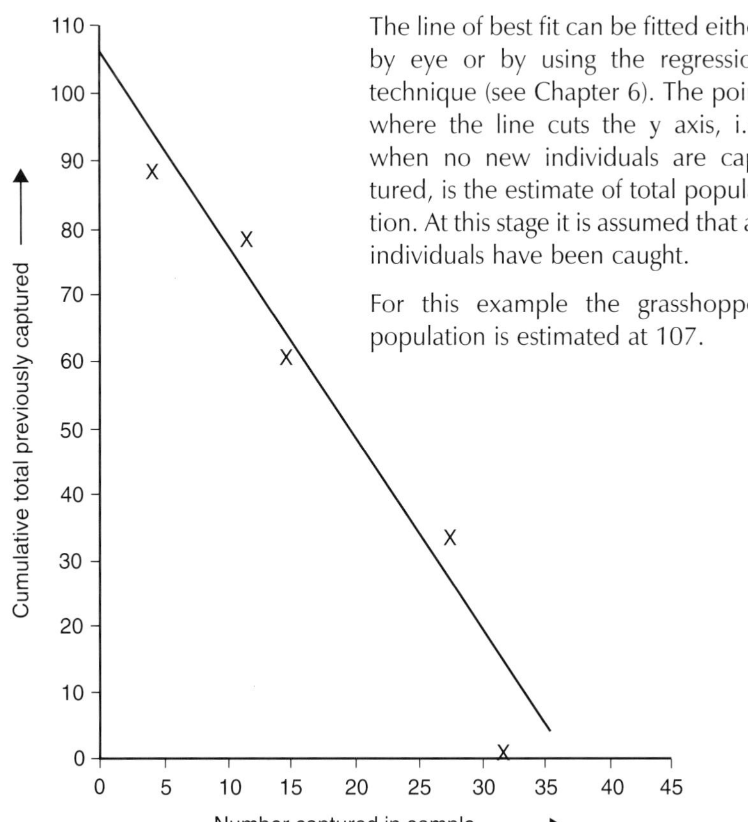

Figure 7.1: Plot of the numbers of grasshoppers caught in each sample against the total number of grasshoppers caught

The line of best fit can be fitted either by eye or by using the regression technique (see Chapter 6). The point where the line cuts the y axis, i.e. when no new individuals are captured, is the estimate of total population. At this stage it is assumed that all individuals have been caught.

For this example the grasshopper population is estimated at 107.

Capture, mark and recapture method

Principles of the method

A sample of the animal species is taken and each individual is marked and released back into the sample area. After an appropriate time interval, the site is sampled a second time and the individuals are recorded as marked or unmarked. The proportion of marked and unmarked individuals provides an estimate of total population numbers.

Calculation

N = estimated number of individuals in the population

a = number of individuals marked

n = *total* number of individuals *captured* in second sample

r = number of *marked* individuals caught in second sample

then $N = \frac{a \cdot n}{r}$

Example

As many sticklebacks as possible were caught by netting from a small pond. These were carefully marked by incising a small nick in the tail with a sharp scalpel. (The method of marking must conform to the assumptions on experimental design as indicated below.)

Sticklebacks were again netted four days later. The second sample should be roughly the same size as the first. The number of marked and unmarked individuals were recorded.

$a = 143$

$n = 129$

$r = 23$

Equation 7.1

$$N = \frac{a \cdot n}{r} = \mathbf{802}$$

The accuracy of this estimate can be established by calculating its **Variance** (σ^2), which is the square of the standard deviation (σ).

Equation 7.2

$$\text{Variance N } (\sigma^2) = \frac{a^2 n(n-r)}{r^3} = \mathbf{22\,982}$$

Equation 7.3

$$\text{Standard deviation of estimate } (\sigma) = \sqrt{22\,982} = \mathbf{152}$$

Referring back to the normal distribution (Chapter 2), we know that approximately 95% of the sample lie within ±2 standard deviations. Therefore we are 95% confident of the estimate of the population size lying in the range 802 ± 304, i.e. between 1106 and 498.

Design of specific survey

1. The more individuals recorded in the survey, the more accurate the result.
2. The method of marking *must not*:
   - harm or injure the individual
   - affect the behaviour of the individual
   - render the individual more likely to be preyed upon.
3. The marking must be sufficiently obvious to the sampler that it is not overlooked if a marked individual is recaptured.
4. The time interval between release and recapture must be great enough to allow random mixing of marked and unmarked individuals but not so long that the mark is likely to be lost.
5. The likelihood of an animal being captured must not change with age.
6. The population should be closed, i.e. no emigration or immigration should occur (unless they can be determined).
7. There should be no births or deaths during the investigation.

8: MEASURING THE DIVERSITY OF COMMUNITIES

A diversity index

In some cases it is useful to quantify the diversity of a community so that either the community may be monitored over a period of time or so that it may be compared with another similar community. The quantity calculated is called the diversity index.

The diversity index is a function of the total number of species and the number of individuals comprising each species. A higher diversity index represents a greater diversity and similarly a lower diversity index represents a lower diversity.

Diversity indices are significant parameters describing a community because they are influenced by the effects of extreme environmental conditions (such as pollution).

It is arguable that the greater the variety of organisms that coexist in a community the more equable are the prevailing conditions. Any extremes of factors such as temperature, exposure, soil pH, oxygen concentrations, presence of pollutants, etc., will tend to make that ecosystem unsuitable for greater numbers of species. By the same argument only a few species will possess the adaptive features to succeed. Consequently a low diversity index will result.

The diversity index is also affected by the successional maturity of the ecosystem.

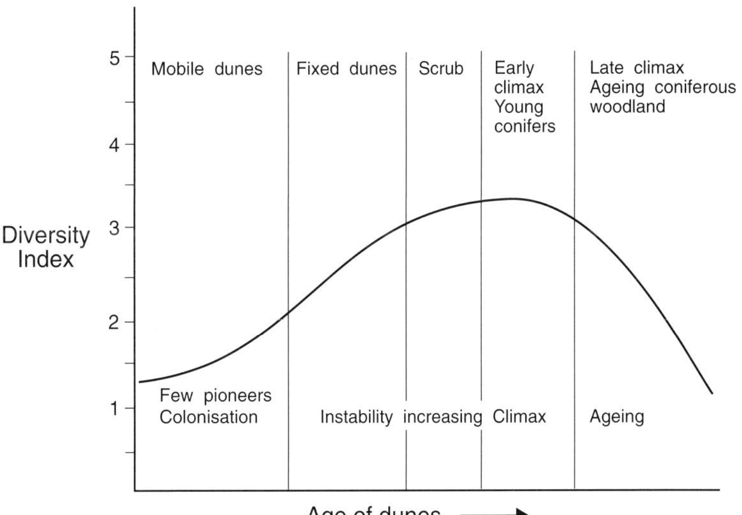

Figure 8.1: Variation of species diversity throughout a sand-dune succession (Ainsdale)

In general terms the diversity of species increases with the development of a succession. At the pioneer stage the substrate to be colonised, e.g. bare sand, may well provide a harsh environment capable of being colonised by few species possessing a high degree of specialisation. As the succession matures, the pioneers modify the environment, rendering it more clement and thus capable of being colonised by an increasing variety of organisms less specialised than their predecessors. The curve sags after the climax stage in some successions where the community may be dominated by few successful individuals, e.g. as coniferous trees reduce the light available to ground flora in the climax stage of a sand-dune succession.

There are several formulae available for calculation of diversity, each having its own advantages and disadvantages. The Simpson Yale index of diversity is illustrated here as it is relatively easy to calculate and hence commonly used.

Equation 8.1

$$D = \frac{N(N-1)}{\sum n(n-1)}$$

where D = Simpson Yale index of diversity
N = total number of individuals
n = number of individuals per species
Σ = sum of

Example

Two rivers were sampled in an identical manner. In each a series of comparable riffle areas were 'kick sampled' for a standardised time. The river Brock is an unpolluted moorland stream; the river Darwen has similar origins but is subjected to a fair degree of domestic and industrial effluent.

The invertebrates in each sample were identified and counted. The results are shown in Table 8.1:

Note: The species do not necessarily have to be identified by name. They need only be sorted into distinguishable groups.

River Brock		River Darwen	
Caddis fly larva	4	Freshwater shrimp	3
Mayfly nymph (type A)	42	Water hog louse	11
Mayfly nymph (type B)	7	Leech (type A)	11
Mayfly nymph (type C)	2	Leech (type B)	4
Mayfly nymph (type D)	3	Midge larvae	3
Mayfly nymph (type E)	2	Tubifex worms	273
Mayfly nymph (type F)	1		
Mayfly nymph (type G)	1		
Stonefly nymph (type A)	4		
Stonefly nymph (type B)	1		
Water beetles	2		
Freshwater shrimp	1		
Total of organisms =	70		305
Total of species =	12		6

Table 8.1: Samples taken from the river Brock and the river Darwen

For the river Brock:

$$D = \frac{N(N-1)}{\sum n(n-1)}$$

$$D = \frac{70 \times 69}{(4\times3)+(42\times41)+(7\times6)+(2\times1)+(3\times2)+(2\times1)+(4\times3)+(2\times1)}$$

$$= \mathbf{2.69}$$

Equation 8.2

For the river Darwen:

$$D = \frac{N(N-1)}{\sum n(n-1)}$$

$$D = \frac{305 \times 304}{(3\times2)+(11\times10)+(11\times10)+(4\times3)+(3\times2)+(273\times272)}$$

$$= \mathbf{1.24}$$

Equation 8.3

The cleaner waters of the river Brock support a greater diversity of invertebrates than those of the river Darwen. The reasons for this disparity must be discovered through further experimentation. The diversity index does not allow us to assume that periodic pollution is the cause.

The sequential comparison index

This is a quicker method than the calculation of the diversity index, and the results compare favourably. This method is recommended where many different samples are being analysed. Examples of its useful application include analysing diversity in freshwater invertebrates, shellfish on a rocky shore, vegetation studies and diversity of shops/services or housing styles in towns. If moving animals in a tray of water are to be examined, anaesthetising them with carbon dioxide will slow them up, making counting much easier.

Example

Note: Accurate identification of organisms is not necessary for the tests of diversity. It is sufficient simply to distinguish that the species are different.

A sample of freshwater invertebrates is taken using a standardised method. The organisms should be randomised by gently stirring, and if necessary anaesthetised. Gently arrange the organisms into lines with a paint brush and record each organism in order.

If, for example, there were four organisms W, X, Y, Z in the sample, the results may be recorded as follows:

ZZZ YZW WWY YZX XXW WXZ Organisms

Each time an organism differs from its immediate predecessor, a new run is initiated:

ZZZ	YZW	WWY	YZX	XXW	WXZ	Organisms
|	| | |	|	| |	|	| |	Runs

The recording may be set out as above with a line indicating the start of each new run.

In this example there are 18 organisms and 10 runs. For statistical validity you must have a minimum of 200 organisms (samples).

The sequential comparison index (SCI) is given by Equation 8.4:

Equation 8.4

$$\text{SCI} = \frac{\text{number of runs}}{\text{number of organisms}}$$

$$= \frac{10}{18} = \mathbf{0.56}$$

Note: accurate identification of organisms is not necessary for the tests of diversity. It is sufficient simply to distinguish that the species are different.

9: MEASURING THE DIVERSITY OF ACTIVITY

The Lorenz Curve

When to use

The Lorenz Curve is a graphical method which is used to display the concentration of activities within an area, for example the degree of industrial specialisation within an urban area. Fieldwork data may be used (frequencies against categories) but it is more common to use secondary sources (e.g. census data).

This technique is particularly useful as it provides a good visual comparison of any observed differences and from it a precise index (the Gini coefficient) can be calculated.

Method

1. First obtain your data. In the worked example on page 43 the data is from the 1871 census returns for Ribchester and Hothersall parishes, Lancashire.
2. Rank the categories and put them in a table in rank order (columns 1 and 2). See Tables 9.1 and 9.2 opposite.
3. Convert each value into a percentage of the total (column 3).
4. Using these percentages, calculate a running total i.e. a cumulative percentage (column 4).
5. On graph paper plot column 2 (horizontal axis) against column 4 (vertical axis).
6. Put on the 'Even distribution line'. This is the line that would be drawn if all the categories were the same size. It runs from the apex of the two axes to the top right-hand corner of the graph (see Figure 9.1).

Example

The same table structure is used for both parishes, but the occupations are in a different order according to rank. This means that they are plotted on the graph not in occupational order but in rank order. **Note:** The use of percentages equalises any difference in the two population sizes.

Note: The use of percentages equalises any difference in the two population sizes.

If we wanted to show a simple contrast between the employment structure of the two parishes we could draw two bar graphs. As it is, we are mainly concerned with deviation from the even distribution, and a contrast between the degrees of concentration of the two parishes can therefore be highlighted.

If the same number of people were employed in each category then the plotted figures would correspond to the line of even distribution. The greater the deviation from this diagonal line, the greater the degree of concentration of one or two occupations in the locality.

As can be seen from our graph (Figure 9.1), Ribchester and Hothersall both deviate from the diagonal showing that both have some degree of concentration of activity. It is Ribchester that displays the greater deviation, and if we refer back to the raw data (Table 9.1) we see that this is caused by the concentration of employment in the textile industries. In Hothersall the deviation is less marked and is mainly due to the predominance of agricultural occupations.

Table 9.1: Employment in the parish of Ribchester — 1871

Occupations	Number Employed 1	Rank 2	Percentage of Total 3	Cumulative Percentage 4
Weavers	195	1	60.0	60.0
Other Textiles	34	2	10.5	70.5
Farm Labourers	25	3	7.6	78.1
Shop Keepers	23	4	7.1	85.2
Tradesmen	18	5	5.5	90.7
Farm Owners	16	6	4.8	95.5
Servants	7	7	2.1	97.6
Professionals	6	8	1.8	99.4
Unemployed	1	9	0.6	100.0

Table 9.2: Employment in the parish of Hothersall — 1871

Occupations	Number Employed 1	Rank 2	Percentage of Total 3	Cumulative Percentage 4
Farm Labourers	19	1	26.3	26.3
Farm Owners	18	2	25.0	51.3
Weavers	14	3	19.4	70.7
Unemployed	6	4	8.3	79.0
Servants	5	5	6.9	85.9
Other Textiles	4	6	5.5	91.4
Tradesmen	3	7	4.2	95.6
Professionals	2	8	2.9	98.5
Shop Keepers	1	9	1.5	100.0

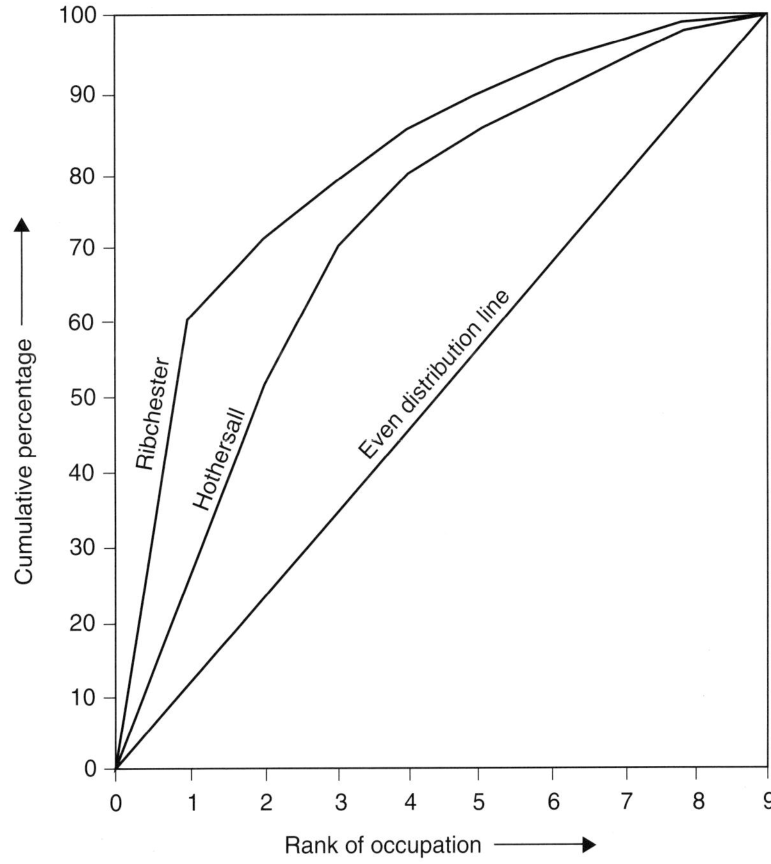

Figure 9.1: Lorenz Curve to show employment structure in the parishes of Ribchester and Hothersall 1871

In conclusion

Lorenz Curves are useful for:

1 Giving a visual impression of the concentration/diversity of activities of a certain locality.

2 Showing differences of such factors between different places. This can be shown on one graph.

3 Showing changes for one location over a period of time.

The main drawbacks are:

1 Care must be taken in devising the categories. If they are too broad then the curve will exaggerate the degree of concentration.

2 They only give a visual impression — for a numerical value we have to analyse the graph further and use the Gini coefficient.

The Gini coefficient

This is easily calculated using the following formula (Equation 9.1):

Equation 9.1

$$\text{Gini coefficient} = \frac{\text{Area of graph above the Even Distribution line (ED)}}{\text{Area of graph between Even Distribution line and the Lorenz curve (AD)}}$$

Finding these areas is best achieved by counting squares on the graph paper and portions of bisected squares. The area above the Even Distribution line is simply half the total area of the graph. The resulting Gini coefficient can range from 1 to infinity. The smaller the number the greater the deviation from the even distribution, therefore the greater the concentration of activities. The larger the coefficient the greater the diversity of activities.

Example

Using the graph previously constructed (Figure 9.1)

Ribchester

$$\text{Gini coefficient} = \frac{\text{ED}}{\text{AD}}$$

Total area of graph = 360 squares

$$\text{therefore ED} = \frac{360}{2} = 180 \text{ squares}$$

AD = 107 squares

Equation 9.2

$$\text{therefore Gini coefficient} = \frac{180}{107} = \mathbf{1.68}$$

Hothersall

ED = 180 (same graph)

AD = 76 squares

Equation 9.3

$$\text{Gini coefficient} = \frac{180}{76} = \mathbf{2.37}$$

These figures confirm our observations from the graph. Further investigations into other parishes in 1871, or the same parishes over time, would be useful.

The Gini coefficient has the same weaknesses as the Lorenz Curve, as the data for its calculation is derived directly from the Lorenz Curve. Despite this, it is a more accurate indicator: it may be difficult to compare two similar Lorenz Curves visually, but the Gini coefficient will give the difference an exact numerical value. With a second variable, e.g. time, correlations may be made using the resulting coefficients.

10: MEASURING DISPERSIONS

When to use

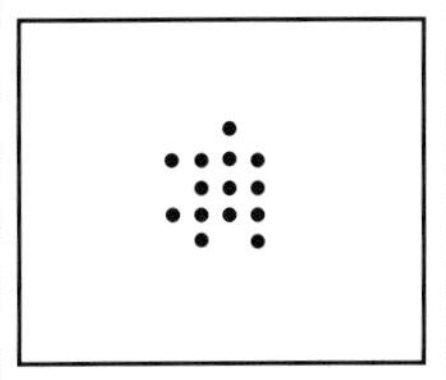

NNI = 0

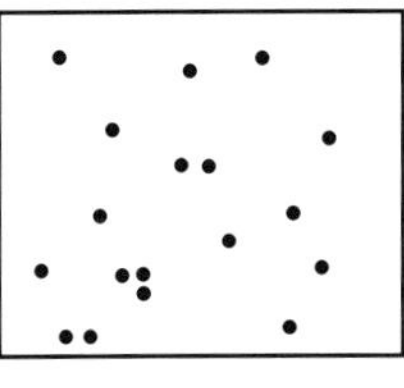

NNI = 1.0

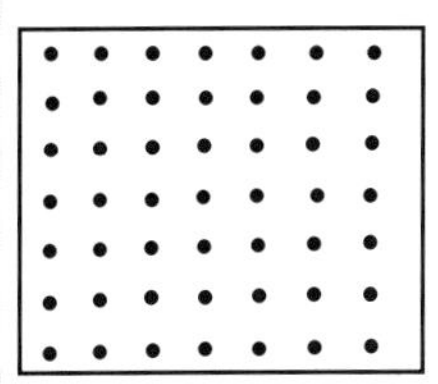

NNI = 2.15

Figure 10.1: Nearest Neighbour Indices

Many fieldwork enquiries involve the observation of spatially arranged data, such as the location of specific functions within an urban area. The usual way of displaying such data is by the use of maps. Maps alone, however, will not describe, interpret or compare patterns numerically, and to do this specific statistical methods have to be adopted.

This chapter deals with two such methods: the nearest neighbour analysis, and quartiles, inter-quartiles and percentiles.

The nearest neighbour analysis

The nearest neighbour analysis provides us with an index. This Nearest Neighbour Index (NNI) provides a test for 'non-randomness' and helps to give a statistical meaning to such terms as 'clustered', 'dispersed', 'random' and to regular distribution of phenomena over space, e.g. villages, functions within the CBD, drumlins, shakeholes.

The index ranges are as follows:

0 = points are completely clustered together

1.0 = points have a completely random distribution

2.15 = points have a completely uniform distribution (points are spread as far apart from each other as is possible)

Figure 10.1 gives a visual impression of these indices.

The Nearest Neighbour Index is usually referred to as R; the formula used to calculate the NNI is given in Equation 10.1:

Equation 10.1

$$R = 2\overline{D}\sqrt{\frac{n}{A}}$$

where $\overline{D}$ = mean observed nearest neighbour distance

n = total number of points in survey

A = area of study

Method

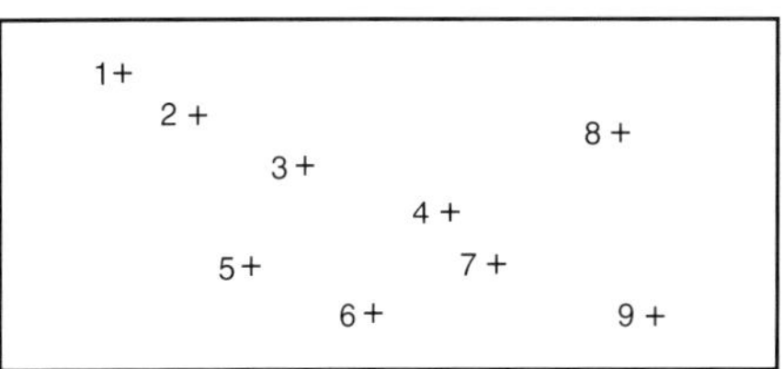

n = 9 A = 36 cm²

Figure 10.2: Newsagents in Burnley CBD

1. Define the boundary of the study area (take care). Calculate its area (A).
2. Plot the points onto the base map and give each one a number (Figure 10.2)
3. Draw up a table as shown in Table 10.1. Measure accurately the distance from each point to its nearest neighbour, i.e. the nearest other point to it. Record this on the table. (It is possible that one point may be the nearest neighbour of several other points — this does not matter.)
4. Calculate the mean of these distances ($\overline{D}$) by adding up all the distances (ΣD) and dividing by the total number of points (n).

Equation 10.2

$$\overline{D} = \frac{\Sigma D}{n}$$

5. Calculate the Nearest Neighbour Index (NNI) using the formula. Take care with units.
6. Relate the answer to the scale of values (see Appendix 6). Does it bear out the initial visual inspection?

Example

The following data, shown in Table 10.1 (distribution of newsagents in Burnley CBD), is also used in the section on the inter-quartile area (see page 48).

Table 10.1

Point number on map	Number of the Nearest Neighbour	Distance between the two points (cm)
1	2	1.1
2	1	1.1
3	2	1.3
4	7	0.4
5	3	1.2
6	7	1.0
7	4	0.4
8	9	2.0
9	8	2.0
		Total (ΣD) = 10.5

The calculations for this example are as follows:

Equation 10.3

$$\overline{D} = \frac{\Sigma D}{n}$$

$$= \frac{10.5}{9} = \mathbf{1.17}$$

The area (A) is 36 cm^2.

Equation 10.4

$$R = 2\overline{D}\sqrt{\frac{n}{A}}$$

$$= 2.34 \times \sqrt{\frac{9}{36}} = 2.34 \times 0.5$$

$$= \mathbf{1.17}$$

As can be seen from our worked example the points are closer to a random distribution than any other type of distribution. By reference to significance tables (Appendix 7) we can confirm whether significant clustering or dispersal has occurred. This is the case when n = 9.

A value less than 0.173 is considered to indicate significant clustering at p = 0.05. Similarly a value greater than 1.287 indicates significant dispersal (p = 0.05).

Comments

At its best this method is used to compare and contrast several distributions rather than just one as in the worked example. For example, drumlin or shakehole distribution in two distinct areas, different CBD functions.

There are several very important problems associated with this method:

1 It cannot always distinguish between a single and a multi-clustered distribution. The two diagrams in Figure 10.3 (left) have an almost identical R value but are visually distributed differently.

2 An index (R value) of 1.0 does not always mean a totally random distribution. Two sub-patterns on the one map when combined in one index may give a totally false impression of randomness, as shown in Figure 10.4 (page 47).

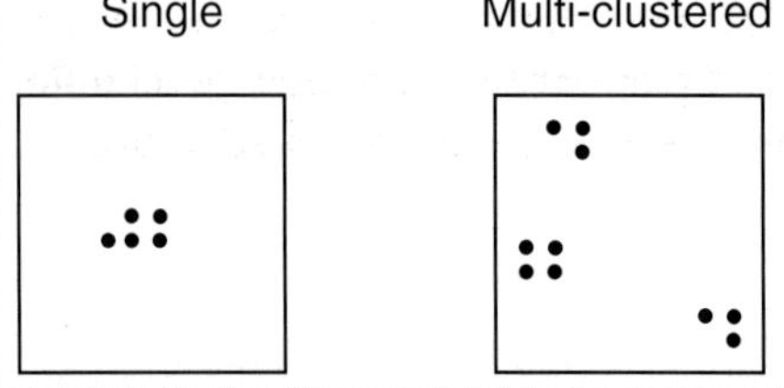

Figure 10.3: Single and multi-clustered distributions which may give an identical Nearest Neighbour Index

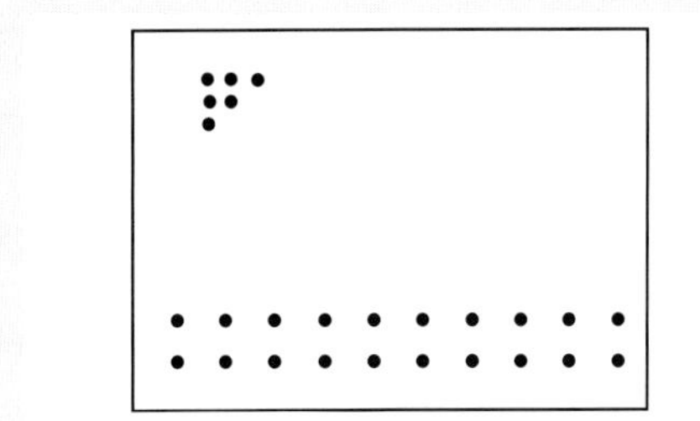

Figure 10.4: Two types of pattern on the one map

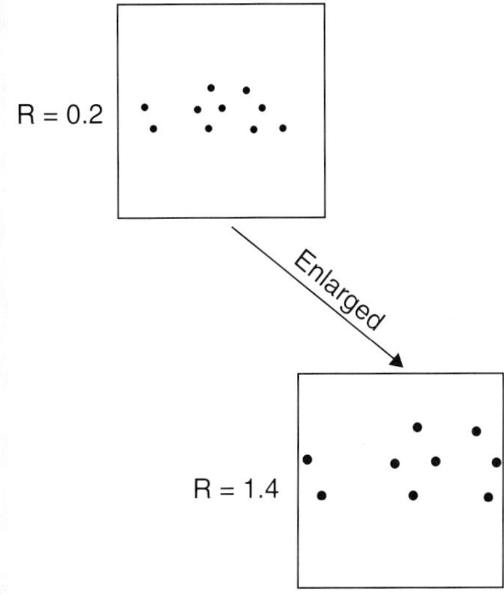

Figure 10.5: The same patterns at different scales

3 Sometimes a value of 1.0 is not caused by the chance distribution of the points mapped; it may be related to a second, unmapped factor. For example, villages in an area may give a value of 1.0 but on closer examination it may be seen that they are all based around springs. It is therefore likely that it is the random distribution of the springs that has caused the R value, and not chance.

4 The NNI obtained may well depend on the area or scale chosen for the map. The two maps in Figure 10.5 (below left) have the same points marked but are of different scales. On one the points appear clustered, whilst on the other they appear dispersed. Note the different R values. As a general rule, when comparing two or more distinct populations use the same scale for the separate base maps, and wherever possible choose boundaries that will give the same sample areas (A) for each.

In conclusion it must be said that great care should be taken in using this method and in choosing and plotting observations. In some cases the inter-quartile method can be a better alternative (page 48).

Quartiles, inter-quartiles and percentiles

We have already seen how to identify the central tendency of data and how to demonstrate how wide ranging such data can be. **We saw that the standard deviation indicates the spread or cluster of data about the mean. Quartiles and percentiles are used to perform a similar function around the median value.**

The inter-percentile range

This is a simple and quickly obtained measure of dispersion, but is a much coarser measure than the standard deviation.

Method

1 The data is arranged in rank order and divided into **four** parts each containing an equal number of values. It is then said to be divided into **quartiles**.

2 The quartile containing the highest values is called the **upper quartile** and the one with the lowest values is the **lower quartile**.

3 The **upper quartile value** is the mean of the lowest value in the upper quartile and the highest value in the quartile below it.

4 Similarly the **lower quartile value** is the mean of the highest value in the lower quartile and the lowest value in the quartile above it.

5 The difference between the upper and lower quartile values is called the **inter-quartile range**. The lower its value the less dispersed is the data.

6 The listings may be divided into any number of equal parts each containing the same percentage of the number of values. In this case the categories are called **percentiles**. The **inter-percentile range** is found by the same method as that described above.

Example

The 'a' axes of 16 pebbles measured on the storm beach at Heysham Head, Lancashire are tabulated in Figure 10.6.

23.7
21.8
21.6
19.9

Upper quartile

Upper quartile value = $\frac{19.9+18.4}{2}$ = **19.15**

18.4
18.0
16.3
15.9

14.9
12.6
10.1
9.4

Lower quartile value = $\frac{9.4+8.4}{2}$ = **8.9**

8.4
7.2
4.3
2.1

Lower quartile

Therefore the **inter-quartile range** = 19.15 – 8.9 = **10.25**

Figure 10.6

This data can now be used in conjunction with other beach data to assess degrees of dispersal of stones throughout the beach.

The inter-quartile area (Q)

A variation of the quartile principle can be used to illustrate the distribution of phenomena spatially: for example, the clustering or dispersal of rural settlements, particularly CBD functions, drumlin and/or shakehole distribution. **It serves a similar function to the nearest neighbour analysis described on page 45. It is quicker and easier to operate, although the end result is usually less precise, but it does overcome some of the problems the NNI faces** (see pages 46–47). The same example is used here as that used in the nearest neighbour analysis.

Method

Note: There must be a minimum of 8 observed points for this method to work.

1 There has to be a **minimum** of 8 observed points for this method to work. Mark their location onto a base map and define the boundaries as you would for the nearest neighbour analysis (page 45).

2 Count up the total number of points (n). Calculate the midpoint: if the number is even then the midpoint is halfway between point n/2 and the next point; if the number is odd than the midpoint is at point (n + 1)/2.

In the case illustrated in Figure 10.7 (page 49) there are 9, so the midpoint comes at point 5.

3 Draw a line vertically through the midpoint counting from the left. Now draw a horizontal line through the midpoint counting down from the top. **Note:** These two midpoints may not be the same.

There should now be an equal number of points in the two vertical halves of the base map and an equal number of points in the two horizontal halves. (In Figure 10.7 there are 4 points on either side of the vertical division and 4 points on either side of the horizontal division.)

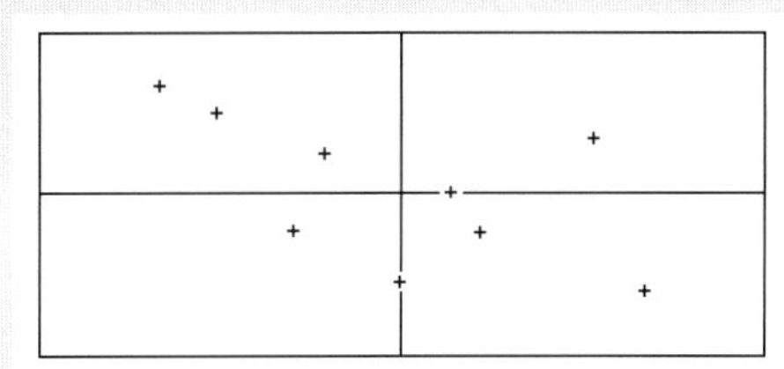

Figure 10.7: Lines drawn through the vertical and horizontal midpoints

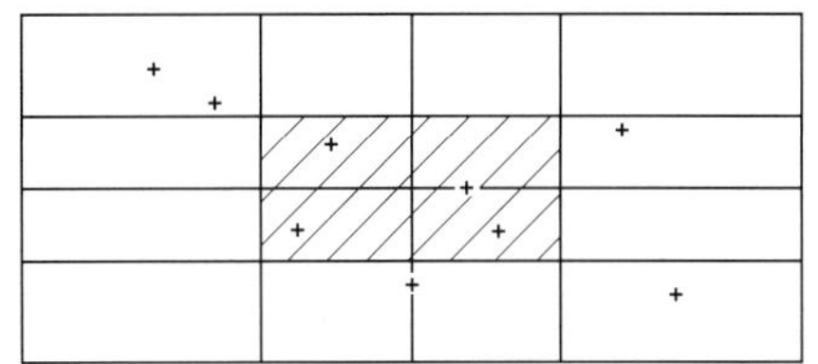

Figure 10.8: Marking the quartiles

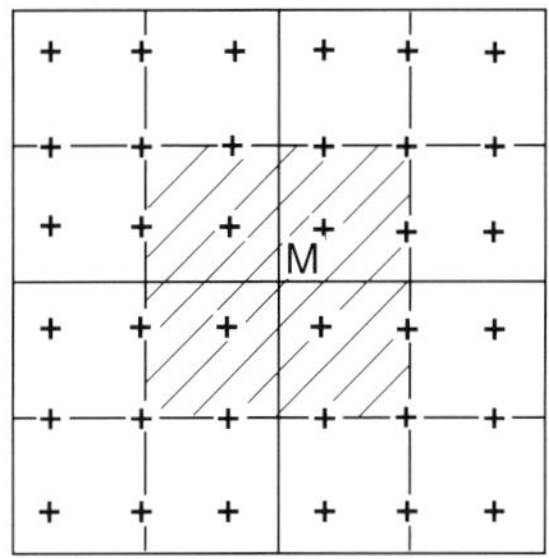

Figure 10.9: Uniform distribution — Id = 0.25

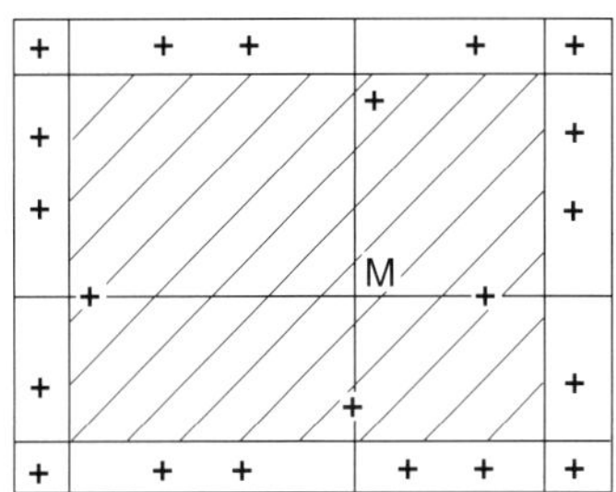

Figure 10.10: Peripheral distribution — Id = 0.61

4 Using the same principle for determining midpoints, divide the map up into 16ths by drawing two more lines vertically and two more lines horizontally (Figure 10.8). In each vertically determined quarter there are now equal number of points (in this example 2) and the same applies to each horizontally determined quarter. These are called **quartiles**.

5 The **inter-quartile area** (Q) is the area of the rectangle formed in the centre of the base map. By careful measurement of the length and breadth the area is easily determined. A small area suggests clustering whilst a large area suggests dispersal. In this case:

Inter-quartile area (Q) (shaded) = 3.5 ˘ 1.5
= **5.25**

6 The resulting figure is only of use when used in conjunction with other similarly derived figures. These can then be used for comparing or contrasting dispersions: for example, several predetermined CBD functions may be mapped to try to determine which tend towards a clustered distribution and which tend towards a dispersed distribution.

7 It is possible from this method to derive a simple index of dispersal (Id). This index relates to the dispersal of points in relation to a central point.

0 = maximum concentration (clustering) around the central point

1 = maximum dispersal from the central point.

The formula is:

$$\text{Id} = \frac{Q}{A}$$

where Q = the inter-quartile area
A = the total area of the unit containing the dispersal.

Our worked example becomes:

$$\text{Id} = \frac{5.25}{36} = \mathbf{0.146}$$

Figures 10.9 and 10.10 show examples for a uniform and for a peripheral point distribution. A uniform distribution gives an Id of 0.25 whilst a peripheral distribution gives a value nearer to 1.

This method is particularly useful when you want to show the dispersal or distribution of features around a central point, e.g. the peak land-value index of a CBD, or villages around a market centre.

11: MEASURING ORIENTATIONS

When to use

One technique of analysing orientated data has already been shown in Chapter 4 using the Chi-squared (χ^2) test. There are, however, tests that have been specifically designed for data that has been expressed as an orientation, e.g. pebbles, shakeholes, drumlins, or wind directions.

Vector analysis

The lines of orientation are often called **vectors**. These vectors have a direction and a magnitude, i.e. **vector direction** and **vector strength**. The purpose of this technique is to show statistically the extent to which the data shows a preferred orientation. Once the preferred orientation has been identified, the relative significance may be shown using significance tables (Appendix 8).

Example 1

The data used in this worked example refers to corrie orientation in North Wales, the same data as that used in Chapter 4 (pages 24–25), but as can be seen it is organised very differently.

1 The data's orientation is put into classes, e.g. 1–40°, 41–80°, 81–120°. These classes are called **azimuths**. The midpoint of each azimuth is then calculated, e.g. for 1–40° the midpoint is 20°. This is called **the midpoint azimuth**. A preliminary table is set up (see Table 11.1) showing the vectors and their frequencies of observation (f).

Midpoint azimuth (ϕ)	Frequency — number of occurrences (f)
20	13
60	13
100	5
140	2
180	4
220	1
260	1
300	2
340	9

Table 11.1

2 Calculate the **resultant vector** (Φ); this is the central point of the distribution.

Equation 11.1

$$\Phi = \tan^{-1}\left(\frac{A}{B}\right)$$

where $A = \Sigma(f \cdot \sin\phi)$

and $B = \Sigma(f \cdot \cos\phi)$

This is best worked out using a table as shown in Table 11.2.

Midpoint azimuth ϕ	Frequency f	sin ϕ	f·sin ϕ	cos ϕ	f·cos ϕ
20	13	0.34	4.42	0.94	12.22
60	13	0.87	11.31	0.50	6.50
100	5	0.98	4.90	–0.17	–0.85
140	2	0.64	1.28	–0.77	–1.54
180	4	0.00	0.00	–1.00	–4.00
220	1	–0.64	–0.64	–0.77	–0.77
260	1	–0.98	–0.98	–0.17	–0.17
300	2	–0.87	–1.74	0.50	1.00
340	9	–0.34	–3.06	0.94	8.46
f(n) = 50		Σ(f·sin ϕ) = 15.49		Σ(f·cos ϕ) = 20.85	

Table 11.2

$$A = \Sigma(f \cdot \sin\phi) = 15.49$$

$$B = \Sigma(f \cdot \cos\phi) = 20.85$$

Equation 11.2

$$\Phi = \tan^{-1}\left(\frac{A}{B}\right)$$

$$= \tan^{-1}\left(\frac{15.49}{20.85}\right) = \mathbf{36.61°}$$

Therefore 36.61° is the central point of the corrie orientation.

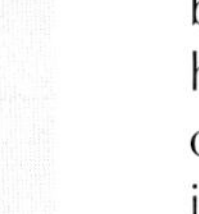

Table 11.2 applies to data that shows a **single orientation** between 1 and 360°; corrie orientation is of this type. Some data, however, may have **two** possible orientations, such as the long axis of a pebble. In this case the formula to be used is slightly different; it now becomes:

Equation 11.3

$$\Phi = 0.5\tan^{-1}\left(\frac{A}{B}\right)$$

where $A = \Sigma(f \cdot \sin 2\phi)$

and $B = \Sigma(f \cdot \cos 2\phi)$

Example 2 on page 52 shows just such a case.

Once the appropriate resultant vector has been calculated then the following stages are identical for both types of orientated data.

3 Calculate the magnitude of the resultant vector (Φ), this is called the **vector strength** (R).

Equation 11.4

$$R = \sqrt{A^2 + B^2}$$

where A and B have the same meaning as before. In this example:

Equation 11.5

$$R = \sqrt{15.49^2 + 20.85^2}$$

$$R = \sqrt{239.94 + 434.72} = \mathbf{25.97}$$

In simple terms:

- If the data gives a large vector strength (R) then the data is clustered around the central point and is likely to show a preferred orientation.
- If the data gives a small vector strength (R) then the data is dispersed throughout the azimuthal ranges, i.e. not just around the central point. In this case the data is not showing a significantly preferred orientation.

4 The vector strength (R) cannot be used on its own as it is dependent on the sample size. The usual practice is to convert it to a percentage value. This is called the **vector magnitude** (L). The formula is:

Equation 11.6

$$L = \frac{R}{n} \times 100$$

where n = total number of observations. In this example:

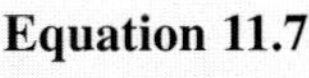

Equation 11.7

$$L = \frac{25.97}{50} \times 100 = \mathbf{51.94\%}$$

By referring to Appendix 8 it can be seen that the point L% 51.94, n = 50 lies above the 0.05 significance line. Thus the assumption may be made that the corries are not orientated at random — the same conclusion that was drawn when this data was analysed using the Chi-squared method (pages 24–25).

Example 2

Vector analysis of semicircular data (pebble orientation)

Mid–point azimuth (ϕ)	2ϕ	sin2ϕ	cos2ϕ	f	f·sin2ϕ	f·cos2ϕ
0	0	0.00	1.00	12	0.00	12.00
20	40	0.64	0.77	16	10.24	12.32
40	80	0.98	0.17	8	7.84	1.36
60	120	0.87	–0.50	4	3.48	–2.00
80	160	0.34	–0.94	3	1.02	–2.82
100	200	–0.34	–0.94	0	0.00	0.00
120	240	–0.87	–0.50	0	0.00	0.00
140	280	–0.98	0.17	3	–2.94	0.51
160	320	–0.64	0.77	4	–2.56	3.08
			Totals:	50	17.08	24.45

Table 11.3: Measuring orientations

$$A = \Sigma(f \cdot \sin 2\phi) = 17.08$$

$$B = \Sigma(f \cdot \cos 2\phi) = 24.45$$

Sample size n = 50 pebbles

$$\text{Resultant vector } \Phi = 0.5\tan^{-1}\left(\frac{A}{B}\right)$$

Equation 11.8

$$= 0.5\tan^{-1}\left(\frac{17.08}{24.45}\right) = \mathbf{17.47°}$$

$$\text{Vector strength } R = \sqrt{A^2 + B^2}$$

Equation 11.9

$$= \sqrt{17.08^2 + 24.45^2} = \mathbf{29.82}$$

$$\text{Vector magnitude } L = \frac{R}{n} \times 100$$

Equation 11.10

$$= \frac{29.82}{50} \times 100 = \mathbf{59.64\%}$$

12: ANALYSING THREE-DIMENSIONAL SHAPES

Zingg diagrams

When to use

These are used when we want to observe three-dimensional shapes such as sediments. The three prime axes are measured (a, b and c).

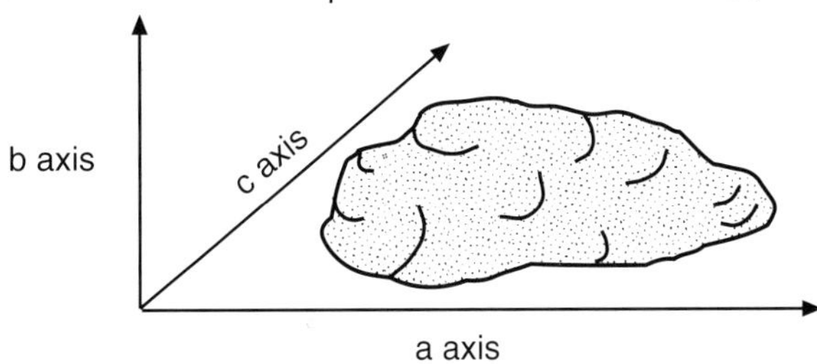

Figure 12.1

From these the following ratios may be determined: b/a and c/b. These ratios should always be less than 1.

Using these two ratios as coordinates, it is possible to classify the observations as one of the shapes shown in Table 12.1.

Shape	b/a	c/b
Spheres	>0.67	>0.67
Discs	>0.67	≤0.67
Rods	≤0.67	>0.67
Blades	≤0.67	≤0.67

Table 12.1

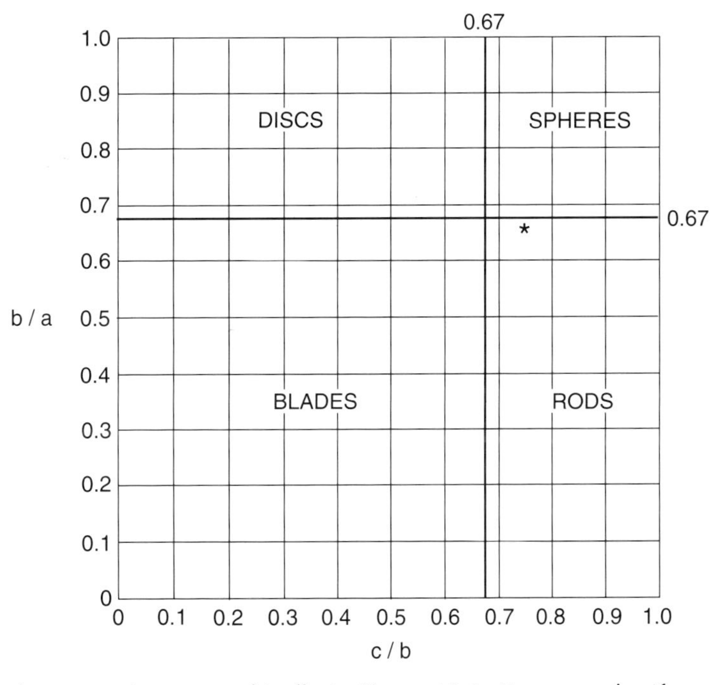

Figure 12.2: Zingg diagram
(* = example in text)

These are shown graphically in Figure 12.2. For example, if a pebble measures 12 cm by 8 cm by 6 cm it would have the following coordinates: b/a = 0.66 and c/b = 0.75. If plotted on the graph it is seen that this particular pebble is classified as a rod.

Using this method different sediments (fluvial, marine, glacial, etc.) can be compared by shape. Equally, different sediments consisting of different rock types may be compared. The percentage of pebbles in each class may be shown by means of bar charts.

The advantage of this diagram is that it gives a quick measure of particle form. Its main disadvantage is that pebbles are assigned to broad categories and are not given a numerical value according to shape. This limits what can be achieved with the results statistically. This can be overcome by using Indices of Sphericity, Flatness and Roundness as devised by Krumbein and Cailleux (see pages 54–55).

Shape indices

In the previous section on Zingg diagrams we saw that by using the a, b and c axes we can classify three-dimensional shapes. As noted, however, the Zingg diagram will not give a numerical value to these shapes. The following section describes the different shape indices that may be used in order to give us this numerical value.

Sphericity (after Krumbein)

Equation 12.1

$$\text{Index of sphericity} = \sqrt[3]{\frac{bc}{a^2}}$$

This index can be shown visually using the same axes as the Zingg classification and the sphericity can be read off this chart in association with the Zingg classes. The values range from a minimum of 0 to a maximum of 1. A true sphere equals 1. In reality most particles fall between 0.3 and 0.9.

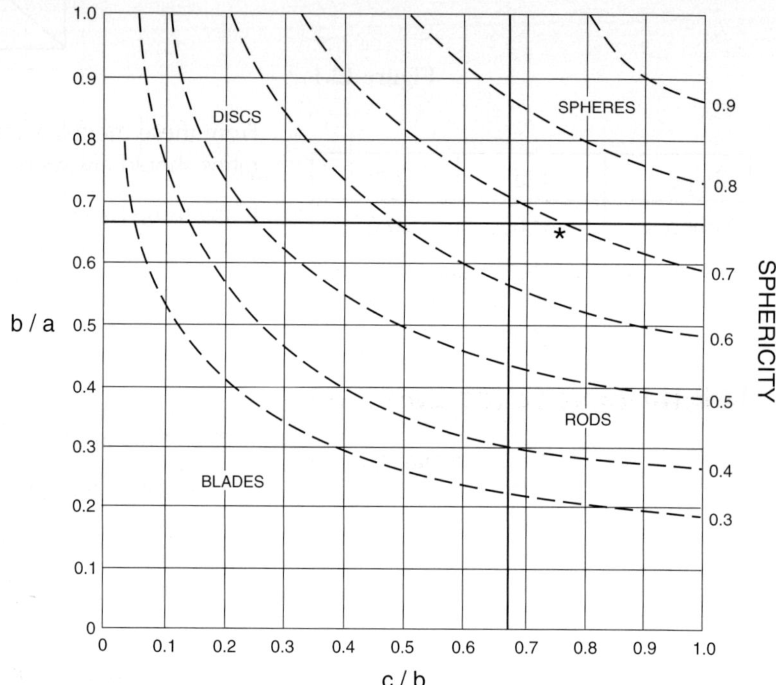

Figure 12.3: Sphericity of three-dimensional shapes

Note: Looking at Figure 12.3, we see that sphericity (Krumbein) and shape (Zingg) are not the same. Particles of a similar sphericity may occur in a wide range of shape classes: for example, a sphericity of 0.5 may be attained by discs, blades or rods. The sphericity is a single value for each particle and as such is suitable for a wide range of statistical analysis.

From our previously worked example we can see from the graph that the sphericity index lies somewhere between 0.6 and 0.7. A more accurate guess using direct measurement off the graph would put the value around 0.69. Using the actual equation we calculate:

Equation 12.2

$$\text{Index of sphericity} = \sqrt[3]{\frac{8 \times 6}{12^2}} = \sqrt[3]{0.33} = \mathbf{0.693}$$

Flatness (after Cailleux)

Equation 12.3

$$\text{Index of flatness} = \frac{a+b}{2c} \times 100$$

In our example:

Equation 12.4

$$\text{Index of flatness} = \frac{12+8}{2 \times 6} \times 100 = \mathbf{166.6}$$

Again the formula is based on the three prime axes. The index ranges from 100 to infinity. In this case the lowest value relates to a perfectly equidimensional shape (a sphere). The flatter the particle, the larger the index, and as such it is essentially the inverse of the sphericity index. Flatness may also be displayed graphically, as in Figure 12.4.

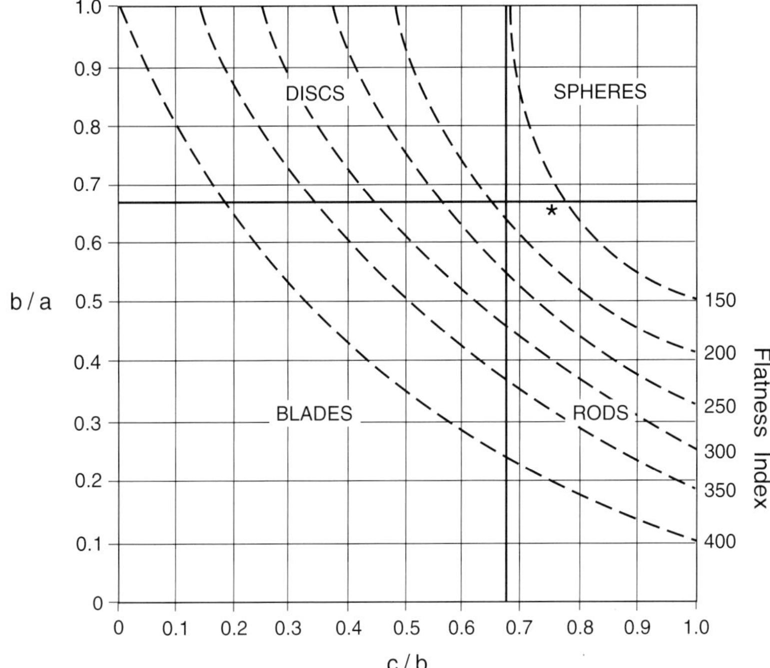

Figure 12.4: Flatness of three-dimensional shapes

Roundness (after Cailleux)

Equation 12.5

$$\text{Index of roundness} = \frac{2r}{a} \times 1000$$

where r = radius of the sharpest corner of the sediment.

This index refers to a two-dimensional shape whereas the other two indices are three-dimensional. The index of roundness is also sometimes referred to as the index of wear. The sharpest/roughest corner (in the plane of maximum projection) is placed upon a series of semicircles of known radius so that the corner just encloses the appropriate semicircle. The radius of the circle (r) is then read and the a axis measured.

The index produces values ranging from 0 to 1000. A perfect circle has a value of 1000.

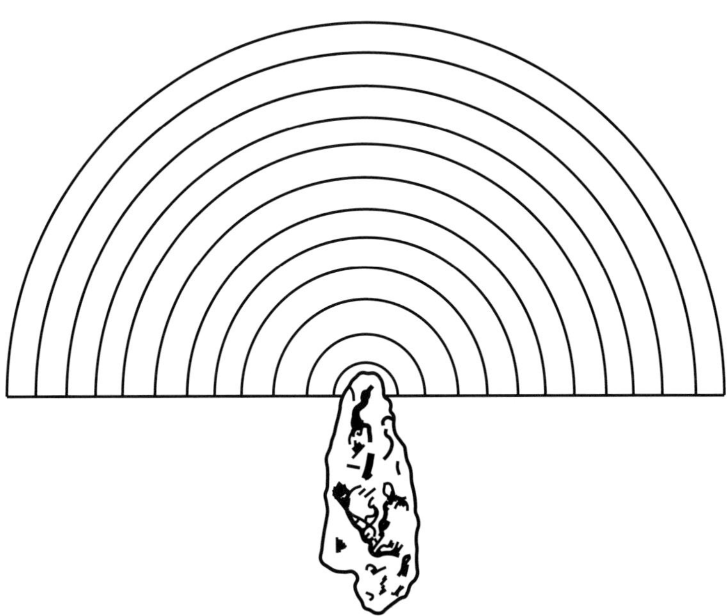

Figure 12.5: Measuring the radius of the sharpest point of a stone using concentric circles

APPENDICES

1: Table of Z-values

The Z-score of a variable is the number of standard deviations by which that variable is above, or below, the mean.

Z	A	B
±0.0	0.000	0.500
±0.1	0.040	0.460
±0.2	0.079	0.421
±0.3	0.119	0.382
±0.4	0.155	0.345
±0.5	0.191	0.309
±0.6	0.226	0.274
±0.7	0.258	0.242
±0.8	0.288	0.212
±0.9	0.316	0.184
±1.0	0.341	0.159
±1.1	0.364	0.136
±1.2	0.385	0.115
±1.3	0.403	0.097
±1.4	0.419	0.081
±1.5	0.433	0.067
±1.6	0.445	0.055
±1.7	0.455	0.045
±1.8	0.464	0.036
±1.9	0.470	0.029
±2.0	0.477	0.023
±2.1	0.482	0.018
±2.2	0.486	0.014
±2.3	0.489	0.011
±2.4	0.492	0.008
±2.5	0.494	0.006
±2.6	0.495	0.006
±2.7	0.496	0.004
±2.8	0.497	0.003
±2.9	0.498	0.002
±3.0	0.499	0.001
±3.1	0.499	0.001
±3.2	0.499	0.001
±3.3	0.500	0.000
±3.4	0.500	0.000

Column A = the probability of a value lying between the mean and the corresponding value of Z.

Column B = the probability of a value exceeding the given value of Z.

2: Critical values of Student's t

Degrees of freedom	Significance level	
	0.05	0.01
1	6.31	63.66
2	2.92	9.93
3	2.35	5.84
4	2.13	4.60
5	2.00	4.03
6	1.94	3.71
7	1.89	3.50
8	1.86	3.36
9	1.83	3.25
10	1.81	3.17
11	1.80	3.11
12	1.78	3.06
13	1.77	3.01
14	1.76	2.98
15	1.75	2.95
16	1.75	2.92
17	1.74	2.90
18	1.73	2.88
19	1.73	2.86
20	1.73	2.85
21	1.72	2.83
22	1.72	2.82
23	1.71	2.81
24	1.71	2.80
25	1.71	2.79
26	1.71	2.78
27	1.70	2.77
28	1.70	2.76
29	1.70	2.76
30	1.70	2.75

To calculate degrees of freedom where the two sample sizes are A and B respectively, DF = (A – 1) + (B – 1).

Reject H_0 if the calculated value of t is greater than the critical value at the chosen significance level.

3: The Mann-Whitney U-test

Critical values of U at p = 0.05

Reject H_0 if the calculated value of U is equal to or less than the appropriate critical value.

n_1	1	2	3	4	5	6	7	8	9	10	11	12	13	14	15	16	17	18	19	20
1	—	—	—	—	—	—	—	—	—	—	—	—	—	—	—	—	—	—	—	—
2	—	—	—	—	—	—	—	0	0	0	0	1	1	1	1	1	2	2	2	2
3	—	—	—	—	0	1	1	2	2	3	3	4	4	5	5	6	6	7	7	8
4	—	—	—	0	1	2	3	4	4	5	6	7	8	9	10	11	11	12	13	13
5	—	—	0	1	2	3	5	6	7	8	9	11	12	13	14	15	17	18	19	20
6	—	—	1	2	3	5	6	8	10	11	13	14	16	17	19	21	22	24	25	27
7	—	—	1	3	5	6	8	10	12	14	16	18	20	22	24	26	28	30	32	34
8	—	0	2	4	6	8	10	13	15	17	19	22	24	26	29	31	34	36	38	41
9	—	0	2	4	7	10	12	15	17	20	23	26	28	31	34	37	39	42	45	48
10	—	0	3	5	8	11	14	17	20	23	26	29	33	36	39	42	45	48	52	55
11	—	0	3	6	9	13	16	19	23	26	30	33	37	40	44	47	51	55	58	62
12	—	1	4	7	11	14	18	22	26	29	33	37	41	45	49	53	57	61	65	69
13	—	1	4	8	12	16	20	24	28	33	37	41	45	50	54	59	63	67	72	76
14	—	1	5	9	13	17	22	26	31	36	40	45	50	55	59	64	67	74	78	83
15	—	1	5	10	14	19	24	29	34	39	44	49	54	59	64	70	75	80	85	90
16	—	1	6	11	15	21	26	31	37	42	47	53	59	64	70	75	81	86	92	98
17	—	2	6	11	17	22	28	34	39	45	51	57	63	67	75	81	87	93	99	105
18	—	2	7	12	18	24	30	36	42	48	55	61	67	74	80	86	93	99	106	112
19	—	2	7	13	19	25	32	38	45	52	58	65	72	78	85	92	99	106	113	119
20	—	2	8	13	20	27	34	41	48	55	62	69	76	83	90	98	105	112	119	127

4: Critical values of Chi-squared

Degrees of freedom	Significance level	
	0.05	0.01
1	3.84	6.64
2	5.99	9.21
3	7.82	11.34
4	9.49	13.28
5	11.08	15.09
6	12.59	16.81
7	14.07	18.48
8	15.51	20.09
9	16.92	21.67
10	18.31	23.21
11	19.68	24.72
12	21.03	26.22
13	22.36	27.69
14	23.68	29.14
15	25.00	30.58
16	26.30	32.00
17	27.59	33.41
18	28.87	34.80
19	30.14	36.19
20	37.57	37.57
21	32.67	38.93
22	33.92	40.29
23	35.18	41.64
24	36.43	42.98
25	37.65	44.31
26	35.88	45.64
27	40.11	46.96
28	41.34	48.28
29	42.56	45.59
30	43.77	50.89
40	55.76	63.69
50	67.51	76.15
60	79.08	88.38
70	90.53	100.43
80	101.88	112.33
90	113.15	124.12
100	124.34	135.81

To calculate degrees of freedom where there are A rows and B columns respectively, use DF = (A – 1) ˘ (B – 1). If there is only one row then there are (B – 1) degrees of freedom.

Reject H_0 if the calculated value of Chi-squared is greater than the critical value at the chosen significance level.

5: Critical values of Spearman Rank Correlation Coefficient

Degrees of freedom	Significance level	
	0.05	0.01
4	1.000	
5	0.900	1.000
6	0.829	0.943
7	0.714	0.893
8	0.643	0.833
9	0.600	0.783
10	0.564	0.745
11	0.523	0.736
12	0.497	0.703
13	0.475	0.673
14	0.457	0.646
15	0.441	0.623
16	0.425	0.601
17	0.412	0.582
18	0.399	0.564
19	0.388	0.549
20	0.377	0.534
21	0.368	0.521
22	0.359	0.508
23	0.351	0.496
24	0.343	0.485
25	0.336	0.475
26	0.329	0.465
27	0.323	0.456
28	0.317	0.448
29	0.311	0.440
30	0.305	0.432

Degrees of freedom = number of paired measurements in total sample.

Reject H_0 if the calculated value exceeds the critical value at the chosen confidence limit.

6: Critical values of Pearson Product-Moment Correlation Coefficient

Degrees of freedom	Significance level	
	0.05	0.01
1	0.9877	0.995
2	0.900	0.980
3	0.805	0.934
4	0.729	0.882
5	0.669	0.833
6	0.622	0.789
7	0.582	0.750
8	0.549	0.716
9	0.521	0.685
10	0.497	0.658
11	0.476	0.634
12	0.458	0.612
13	0.441	0.592
14	0.426	0.574
15	0.412	0.558
16	0.400	0.543
17	0.389	0.529
18	0.378	0.516
19	0.369	0.503
20	0.360	0.492
25	0.323	0.445
30	0.296	0.409
35	0.275	0.381
40	0.257	0.358
45	0.243	0.338
50	0.231	0.322
60	0.211	0.295
70	0.195	0.274
80	0.183	0.257
90	0.173	0.242
100	0.164	0.230

Degrees of freedom = N – 1 where N is the number of paired observations.

Reject H_0 if the calculated value is greater than the critical value (in absolute terms) at the chosen significance level.

7: Critical values of the Nearest Neighbour Index

n	Clustered pattern		Dispersed pattern	
	0.05	0.01	0.05	0.01
2	0.392	0.140	1.608	1.860
3	0.504	0.298	1.497	1.702
4	0.570	0.392	1.430	1.608
5	0.616	0.456	1.385	1.544
6	0.649	0.504	1.351	1.497
7	0.675	0.540	1.325	1.460
8	0.696	0.570	1.304	1.430
9	0.713	0.595	1.287	1.406
10	0.728	0.615	1.272	1.385
11	0.741	0.633	1.259	1.367
12	0.752	0.649	1.248	1.351
13	0.762	0.663	1.239	1.337
14	0.770	0.675	1.230	1.325
15	0.778	0.686	1.222	1.314
16	0.785	0.696	1.215	1.304
17	0.792	0.705	1.209	1.295
18	0.797	0.713	1.203	1.287
19	0.803	0.721	1.197	1.279
20	0.808	0.728	1.192	1.272
21	0.812	0.735	1.188	1.266
22	0.817	0.741	1.183	1.259
23	0.821	0.746	1.179	1.254
24	0.825	0.752	1.176	1.248
25	0.828	0.757	1.172	1.243
26	0.831	0.762	1.169	1.239
27	0.835	0.766	1.166	1.234
28	0.838	0.770	1.163	1.230
29	0.840	0.774	1.160	1.226
30	0.843	0.778	1.157	1.222
31	0.846	0.782	1.155	1.219
32	0.848	0.785	1.152	1.215
33	0.850	0.788	1.150	1.212
34	0.853	0.791	1.148	1.209
35	0.855	0.794	1.145	1.206
36	0.857	0.797	1.143	1.203
37	0.859	0.800	1.141	1.200
38	0.861	0.803	1.140	1.197
39	0.862	0.805	1.138	1.195
40	0.864	0.808	1.136	1.192
41	0.866	0.810	1.134	1.190
42	0.867	0.812	1.133	1.188
43	0.869	0.815	1.131	1.186
44	0.870	0.817	1.130	1.183
45	0.872	0.819	1.128	1.181
50	0.878	0.828	1.122	1.172
60	0.889	0.843	1.111	1.157
70	0.897	0.855	1.103	1.145
80	0.904	0.864	1.096	1.136
90	0.909	0.872	1.091	1.128
100	0.914	0.378	1.086	1.122

n = number of points in the survey.

To test for clustering: reject H_0 if the calculated value of R is less than the critical value at the chosen significance level.

To test for dispersion: reject H_0 if the calculated value of R is greater than the critical value at the chosen significance level.

8: Significance of preferred orientation (from Lord Rayleight)

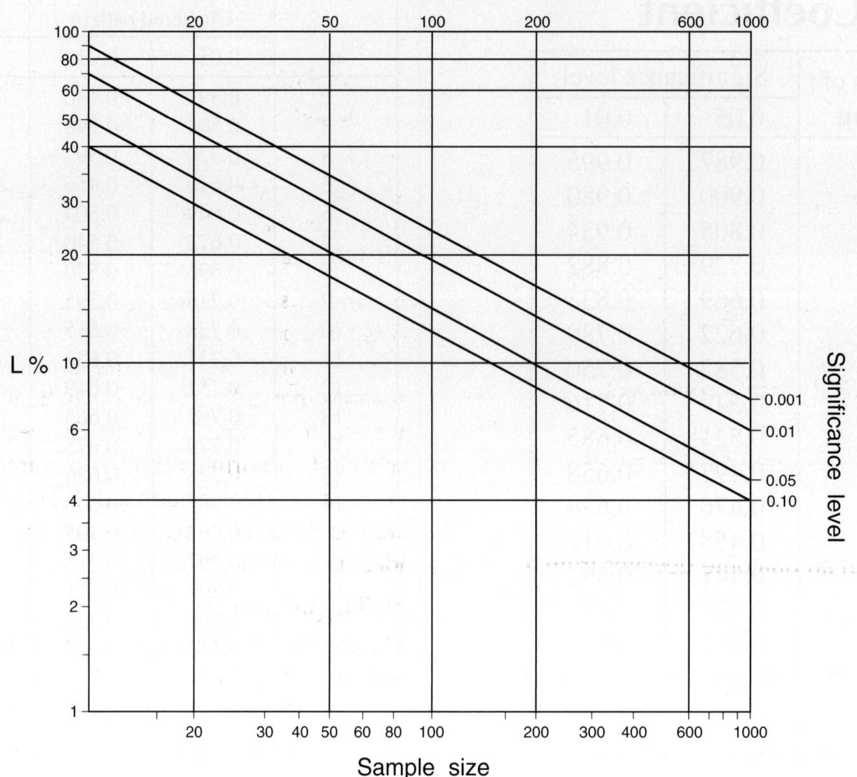

To test for orientation: reject H_0 if the calculated value of L is greater than the value (diagonal line) at chosen significance level.

GLOSSARY OF WORDS AND TERMS

Cross-references within this glossary are indicated by SMALL CAPITALS.

Azimuth
Orientation or bearing in degrees.

Categorical level of data *see* DATA LEVEL — CATEGORICAL

Clustered
A group of items occurring closer together than would be expected from a RANDOM distribution.

Coefficient
In a strict mathematical sense a coefficient is a quantity placed before, and multiplying, another factor. In statistics a coefficient may be synonymous with an INDEX, e.g. coefficient of correlation.

Confidence limits
The degree to which an outcome deviates from a chance occurrence.

Correlation
The strength of the relationship between two values.

Data (plural)
Facts or collected information from which inferences are made.

Data level — categorical
Data which cannot be put into an order, only a category, e.g. woodland plants, marsh plants, moorland plants.

Data level — interval
Measurements which have precise values so that the magnitude of difference between individuals can be calculated, e.g. height, weight, temperature.

Data level — ordinal
Ranked data which does not assess the size of the difference.

Degrees of freedom
A number which represents the size of a sample. The method of calculation varies according to the statistical test used.

Dispersed
A group of items occurring further away from each other than would be expected from a RANDOM distribution.

Diversity
The degree of difference within a sample.

Frequency
The number of times that a particular quantity (or range of quantities) occurs.

Hypothesis (H_1)
An idea which is used as a starting point for an investigation, without the assumption that the idea is correct. *See also* NULL HYPOTHESIS.

Index
A scale of relative values. *See also* COEFFICIENT.

Interval level of data *see* DATA LEVEL — INTERVAL

Mean
The total of a series of quantities divided by the number of quantities in the series.

Median
The value within a series of quantities which lies in the middle of that series, i.e. there are the same number of quantities above as below the median value.

Normal distribution
A bell-shaped curve which is perfectly symmetrical, with the mean and median at the same point. It must be realised that collected data may be insufficient to fulfil the above statement exactly. To use the statistical theory relating to a normal distribution it must be inferred that your data is *from* a normal distribution, i.e. as more data were collected the nearer your curve would tend to the ideal normal distribution.

Null hypothesis (H_0)
The reverse of the HYPOTHESIS. If the hypothesis is that 'A' influences 'B', then the null hypothesis would be of the form that A and B vary independently of each other. In statistical theory it is considered more objective to disprove H_0 than to prove H_1.

Ordinal level of data *see* DATA LEVEL — ORDINAL

Orientation
Position relative to points of the compass.

Probability
The likelihood of a particular event occurring, usually written as p. If $p = 1$ then the event is certain to happen. If $p = 0$ then it is impossible.

Random
The equal chance of any occurrence. If an item is sampled randomly then it has exactly the same chance of being chosen as all other items within the population.

Significance level
The total PROBABILITY of all outcomes contained in a critical region (usually 5%).

Skewed curve
A non-symmetrical distribution bias towards one end of the curve.

Standard deviation
A statistical expression of the degree of dispersion of values in a NORMAL DISTRIBUTION.

Symmetrical curve
A curve which, when divided by a central line, is identical in size and shape either side of that line.

Variance
The square of the STANDARD DEVIATION.

Vector
Quantity having both direction and magnitude, often used to determine the relative positions in space of two points.

REFERENCES

As explained in the Introduction, this book is designed as a recipe book of statistical tests which often prove necessary in the analysis of field-work data. Very little mathematical rationale is explained, and only the more commonly used tests are described. For greater depth of treatment the following books may prove useful.

Bailey, P. and Fox, P. S. (1996) *Geography Teachers' Handbook*, The Geographical Association.

Briggs, D. (1981) *Sediments*, Butterworth.

Chalmers, N. and Parker, P. (1986) *Fieldwork and Statistics for Ecological Projects* (OU Project Guide), Field Studies Council.

Clarke, G. M. and Cooke, D. (1983) *A Basic Course in Statistics,* Arnold.

Dowdeswell, W. H. (1984) *Ecology, Principles and Practice*, Heinemann.

Ebdon, D. (1977) *Statistics in Geography,* Blackwell.

Hammond, R. and McCullagh, P. S. (1978) *Quantitative Techniques in Geography,* Oxford.

Lenon, B. J. and Cleves, P. G. (1984) *Techniques and Fieldwork in Geography,* UTP.

Mathews, J. A. (1981) *Quantitative and Statistical Approaches to Geography*, Pergamon.

Wratten, S. D. and Fry, G. L. A. (1980) *Field and Laboratory Exercises in Ecology,* Arnold.

In addition, the following publication may prove useful in conjunction with this book:

St John, P. R. and Richardson, D. A. (1990) *Methods of Presenting Fieldwork Data*, The Geographical Association.

COMPUTER SOFTWARE

Many of the statistical tests described in this book involve relatively complex formulae and as such may be time-consuming and open to student error. There are now several computer packages available that are both easy to use and will save a large amount of 'number-crunching' time. Some of these are listed below.

Advanced Level Software

Place of Publication:	Cambridge
Publisher:	Cambridge University Press
Telephone:	01223 312393
Year of Publication:	1991
Machine/Version/Price:	BBC B/£95.00+VAT
Format:	6 disks; teacher's handbook
Compatible with:	40/80t
ISBN:	0 52132164 6

Contents: basic probability; binomial; poisson and normal distributions; other probability density functions; sampling distributions; unbiased estimators; the t-distribution and confidence intervals; hypothesis testing using parametric and non-parametric tests; the Chi-squared distribution and Chi-squared tests, correlation; statistics for economics; data disk.

GraphBox Professional

Place of Publication:	Exeter
Publisher:	Minerva Software
Telephone:	01392 437756
Year of Publication:	1991
Machine/Version/Price:	Acorn Archimedes; Acorn A5000 /£129.95 inc VAT
Format:	Floppy disk; manual
Compatible with:	RISC OS

A multitasking graphing package with over 40 graph types and statistical information. Multiple charts; cross-hatch shading; up to 16 times zoom; best-fit line/curve; measures positions; uses outline fonts; produces drawfiles. Hotlink to PipeDream 3. Charts include scientific, mathematical and presentation graphs. Site licences available. Product is specifically geared to meeting the needs of key stage 4.

Maths with a micro: using software

Author:	SMP 11–16
Place of Publication:	Cambridge
Publisher:	Cambridge University Press
Year of Publication:	1990
Series:	SMP 11–16
Machine/Version/Price:	BBC B; BBC B+; BBC Master 128; RM Nimbus/£20.00
Format:	2 disks; 85pp worksheet ringbinder
ISBN:	0 52137881 8

This pack concentrates on ways in which existing software can be used to enhance mathematics teaching and the index shows how available software links with SMP 11–16. It covers many different kinds of software, including investigations, graph plotting, spreadsheet and statistics programs. There are about 10 worksheet masters provided for use with these programs.

Melstat

Place of Publication:	Melton Mowbray
Publisher:	NW Richards (Educational Software)
Telephone:	01664 68335
Year of Publication:	1989
Machine/Version/Price:	RM Nimbus; IBM PC and compatibles/2.9/£25.00
Format:	Floppy disk

Allows user to create and edit a table of raw statistical data. Both numeric and textual data accepted. Will perform: a wide range of statistical computations — mean, variance, etc.; correlation; regression; tabulation; analysis of variance; non-parametric indices. Site licences available.

Statistics

Author:	Sweetens
Place of Publication:	Abingdon
Publisher:	Independent Software
Telephone:	01235 535035
Year of Publication:	1985
Machine/Version/Price:	BBC Master 128/£27.96
Format:	1 ˘ 5¼-inch floppy disk; booklet

The booklet sets out some of the ways the programs can be used both in and out of the classroom. Exercises are suggested for pupils ranging from routine calculations to extended investigations. The programs contain material to enhance the learning process where traditional methods are difficult. Two simulations are included which bridge the gap between the need to use real data and the lack of time to collect it. Site licences available.

Statistics for Social Scientists Software

Place of Publication:	Cambridge
Publisher:	Cambridge University Press
Telephone:	01223 312393
Year of Publication:	1991
Machine/Version/Price:	BBC B/£95.00+VAT
Format:	4 disks; teacher's handbook
Software comes with:	BBC; 40t or 80t disk drive
ISBN:	0 52132163 8

Contents: the Chi-squared distribution and Chi-squared tests (simplified versions); correlations (simplified version); economic statistics; hypothesis testing; data disk.

Supastat (version 3.2)

Place of Publication:	Tewkesbury
Publisher:	Software Production Associates (SPA) Ltd
Telephone:	01684 833700
Year of publication:	1995
Machine/Version/Price:	IBM PC and compatibles; RM Nimbus 186/school suite/ £36.00+VAT/full suite/£72.00+VAT IBM PC network; RM Nimbus network/school suite/£90.00+VAT/ full suite/£190+VAT
Software comes with:	Windows installation included

A data-analysis suite of programs. Full suite gives an extensive set of tests designed for professional use; the school suite is a subset providing sufficient statistical power for use up to and including A-level statistics. Site licences and upgrades available.

WinLang

Author:	Knight, S.
Place of Publication:	Brighton
Publisher:	Trellis Education Software and Training
Telephone:	01273 203920
Year of Publication:	1994
Series:	Win
Machine/Version/Price:	IBM PC compatibles/3/£60.00
Format:	Disk; manual
Software requires (min.):	Microsoft Windows 3.1+; colour monitor; mouse

A departmental assessment package designed to allow easy access to national curriculum statements, statistics reports, profiles and lists. Upgrades to NCSoft4 available. The product is specifically geared to meeting the needs of key stages 1, 2, 3 and 4 covering national curriculum subjects, English and modern languages.

1st

Authors:	Edwards, G.R. and Turnbull, C.
Place of Publication:	Willaston
Publisher:	Serious Statistical Software
Date of Publication:	1992
Machine/Version/Price:	Acorn Archimedes/1.3/£205.00
Format:	2 ˇ floppy disk; manual
Software requires (min.):	1MB RISC OS 2 or 3 machines

Data-analysis and manipulation software covering elementary to advanced statistical techniques. Multi-tasking with spreadsheet-type data entry and CSV file interface to other software. Drawfile graphics. Useful for 'value added' analyses.

Note: It is the opinion of the authors that although computers can be a useful tool in the field of statistical analysis, students should at first attempt the required tests manually. By manually following a test through, the user is more likely to understand the true meaning of the end result. The authors believe that merely feeding data into a machine which then throws out an answer does not lead to a full understanding.

As already shown many of these tests may be performed in table form. Row and column manipulation (totalling, averaging, etc.) are common. Powerful spreadsheets (such as Lotus, Works, Excel) now form an important part in most schools' information technology curriculum. A well-designed spreadsheet will not only carry out these complex calculations automatically but may also illustrate your results using an appropriate graphing technique.

Many commercial databases such as 'Key' will also provide such statistical tests as mean, standard deviation, Spearman Rank and Chi-squared.

'Know then fully the nature of the beast so that you may be its ruler and not its slave.'